Macmillan Building and Surveying Series
Series Editor: Ivor H. Seeley
Emeritus Professor, Nottingham Trent University

Macmillan Building and Surveying Series
Series Standing Order ISBN 0–333–69333–7

You can receive future titles in this series as they are published by placing a
standing order. Please contact your bookseller or, in the case of difficulty, write
to us at the address below with your name and address, the title of the series
and the ISBN quoted above.

Customer Services Department, Macmillan Distribution Ltd
Houndmills, Basingstoke, Hampshire, RG21 6XS, England.

Construction Economics

An Introduction

Stephen L. Gruneberg

Lecturer in Economics
School of Construction
South Bank University
London

MACMILLAN

First published 1997 by
MACMILLAN PRESS LTD
Houndmills, Basingstoke, Hampshire RG21 6XS
and London
Companies and representatives
throughout the world

ISBN 0-333-65541-9

A catalogue record for this book is available
from the British Library.

This book is printed on paper suitable for recycling and made
from fully managed and sustained forest sources.

10 9 8 7 6 5 4 3 2 1
06 05 04 03 02 01 00 99 98 97

Copy-edited and typeset by Povey–Edmondson
Tavistock and Rochdale, England

Printed in Hong Kong

To my mother

Contents

Preface

The aim of this book is to introduce students and practitioners of construction and property to economics, with a view to preparing them for a professional life in which they will inevitably discuss the topics raised here with clients as well as professionals from the financial world. It is essential that the comments and contributions of the planner, developer, architect, engineer, builder and surveyor are well informed and expressed in the vocabulary of colleagues in the other professions. The study of economics will also give them the ability and the confidence to communicate with professionals from other disciplines.

The theoretical tools of economics can be used by members of the construction and property professions and others to enable them to become imaginative and productive providers of the built environment. The study of economics provides an approach to, and an understanding of, the implications of design and construction, not only in an economic but also in a social context.

The study of economics suggests several general principles. It may be worthwhile, even at this stage, to list some of these. This list is not in any order of priority.

1. A correct decision for an individual is not necessarily a correct decision if everybody decides to do the same thing.
2. Decisions are often taken in terms of a little more or a little less.
3. Let bygones be bygones.
4. What matters is the real world; money only matters in so far as it affects the real world.
5. Always shop around before making a decision – consider the options.
6. From a firm's point of view, it is more important to be profitable than efficient.
7. There is no one way of doing anything – the best way depends on the objectives.
8. Nothing in economics is good or bad – it all depends on whose interests you are considering.
9. Everything has a cost; not all costs can be measured in terms of money.
10. Exchange can make people better off, even when no more is produced.

It is well known that economists rarely agree (partly because of the eighth principle), and it is unlikely that there would be universal agreement on this list, gleaned from a study of the subject. There might not even be agreement about the list itself. Nevertheless, it is to be hoped this book will clarify some of these statements. As it is, the book ranges over a wide variety of subjects in order to provide a bird's-eye view of economics in the context of the production of the built environment. It is also hoped that readers will be encouraged by this book to read further and deeper.

After an initial introduction to economics in the first chapter, the book is divided into three parts: the construction sector, the economics of the firm, and project appraisal.

Chapter 1 is concerned with the subject of economics itself and the economy as a whole. It aims to show that the study of economics and the economy as a whole have important lessons for the management of construction projects and property.

Part 1 deals with the economic context of the construction and property sector, providing a brief summary of its development, the range of built output, the types of firm, and their working practices and arrangements. Chapter 2 shows that the current arrangements of the construction industry have largely emerged from historical precedents and traditionally demarcated functions. Chapter 3 examines the role of the construction industry in the economy as a whole, showing the theoretical economic links between construction and its customers in both the private and public sectors.

Chapter 4 deals with these links in more detail, showing how construction markets are conventionally described in terms of supply and demand. Chapter 5 then extends the concept of markets in a construction context to describe the variety of markets that exist there and to show how markets operate within the sector. Finally, the construction labour market is discussed in more detail in Chapter 6. This completes the first part of the book.

In Part 2 the perspective is seen from a firm's point of view. A firm can range in size and type from a professional practice or small contractor employing as few as one or two people to a large contractor employing many hundreds. Chapter 7 looks at the economics of costs and revenues. This equips the reader for the discussion of the conventional theory of market types in Chapter 8. In Chapter 9 revenues, costs and profits are introduced in terms of balance sheets and profit and loss accounts. Chapter 10 concludes Part 2, with a discussion of business planning and project management.

In Part 3, Chapter 11 introduces the general concepts and principles of feasibility studies, while Chapter 12 describes the discounting techniques which lie at the heart of financial investment appraisal. Chapter 13 looks at the wider economic and social issues raised by cost–benefit analysis.

STEPHEN L. GRUNEBERG

Acknowledgements

I should like to thank the many people who have given me their help, advice and comments, all of which have been invaluable to me in the writing of this book. I am therefore very grateful to my friends and colleagues, including John Adriaanse, Michael Ball, Derek Beck, Linda Clarke, Jeff Denton, Janet Druker, Graham Ive, Farzad Khosrowshahi, Julia Lemessany, Keith Povey, James Semple, Malcolm Stewart and Frank Woods, all of whom have given me sound advice and the benefit of their knowledge. I am also very grateful to Norman Allen, of the Plant Hire Association, and Peter Gill, of the Builders Merchants Federation, for their time and help. Above all, my long-suffering wife, Jan, who has supported me throughout, deserves a special thank you.

1 Introduction to the Economics of Construction

Introduction

This chapter defines the subject area of economics and discusses the nature of the basic economic problem. It shows that the study of economics is relevant to those interested in construction.

Many people think of building as a temporary inconvenience, with the work taking place on a construction site for a relatively short period. For example, the apologetic notice often posted up by shops when building work is carried out is, 'Business as usual'. One can only wonder at the usual service such businesses offer! In reality, construction is a permanent feature of any economy. The construction process takes place continually as part of the annual production output of every society. Building and civil engineering work moves from temporary site to temporary site, but the processes of construction, building maintenance and renewal have always been an essential economic activity in every society since people first began to provide themselves with shelter.

The economic background to the building industry in general and the construction professions in particular consists of certain economic concepts and theories. These theories provide the foundation upon which the methods of assessing the economic viability of buildings and the built environment are based. The purpose of this chapter is to place both the construction industry and the professions in an economic context. The chapter therefore begins with an introduction to the theoretical economic framework. Having introduced the reader to a brief description of the subject area of economics in general, the chapter then deals with the particular economic problems and constraints facing architects, quantity surveyors, structural engineers and, of course, contractors. The chapter will conclude with a brief assessment of the size, contribution and cost of the construction industry to the economy as a whole, as well as a brief description of the historical development of the construction professions to show their economic roles *vis-à-vis* the construction industry.

The basic economic problem

The term *economics* is derived from the Greek language and means *house-keeping*. Every household receives an income, and members of a household must decide how best to spend that money between all the goods and services available to meet their conflicting needs. In essence, economics is concerned with the allocation of scarce resources. Having more of one thing implies having less of something else.

The same problem is faced by firms and governments. Firms receive their income from sales and must choose between the investment opportunities open to them. Governments receive their income mainly in the form of tax revenues and must decide which policy options to select. Unfortunately, whether one is concerned with households, firms or governments, there is never enough money to satisfy all the needs that people express.

Doubling the amount of money in existence would be no solution. That would be analogous to doubling the number of shares in a company. Each share would then be worth half its previous value; and the firm would still be the same size. The fact that there is not enough money simply reflects the fact that actual resources are scarce.

Money and the real economy

It is useful at the outset to establish the role of money in the economy and in the building industry. Economists draw attention to the distinction between the real economy and the money economy. The real economy is concerned with the actual goods and services available, which, depending on their distribution throughout the population, determine the standard of living within a community. Thus the number of cars, the quantity of food and the quality of health care are examples of the goods and services in the real economy, as are the number, size, quality and location of dwellings, the accessibility of shopping facilities, the number of hospital beds, theatre seats, libraries, museums and railway stations, and even the kilometres of roads and railways, including bridges and tunnels.

The real economy may be measured by the actual goods and services provided. Money, on the other hand, is used to enable transactions to take place. It is not generally desired for its own sake, but for the produce it can buy. Money savings may provide individuals with a sense of security and independence, because it represents stored wealth to be exchanged for goods in the future. A money value is used as a signal of the relative worth of goods and services in the marketplace. This monetary system obviates the inconvenience of bartering, which requires individuals to find others who possess a desired product and want to exchange it for another on offer.

The question arises, what is money? Money is a liquid asset. Liquidity is the characteristic of an asset which enables it to be converted into other assets without delay or loss of value. Not all assets are liquid: selling property may take months; selling a car may incur marketing and advertising expenses, which the vendor must deduct from the value of the vehicle. Some assets, such as shares in building societies, must be cashed before transactions can be carried out. Because the delay involved in cashing such assets is minimal, they are often referred to as 'near money'.

Money is the most liquid form of asset and represents an individual's purchasing power, which is his or her entitlement to make a claim for a share of the goods and services available in the economy. The purchasing power of the pound has declined continuously this century, as each firm raises its prices because others do, resulting in inflation. Inflation is the rate per annum at which money loses its value.

Money is whatever money does and, generally, it has several functions. The most important function is to act as a *medium of exchange*. This function enables individuals to conduct their affairs by offering their goods, services or labour in return for money, as described previously. Money is also used as a *unit of account* or measure of value to evaluate costs and revenues. It is used to compare the relative costs and prices of components, service, goods and wealth. Money may also be accumulated by saving to form a *store of wealth* provided inflation does not reduce its purchasing power over time. The rate of inflation is the rate at which money loses its purchasing power.

In order to carry out its functions efficiently, money requires several characteristics. It must be acceptable, otherwise transactions cannot take place. It must be recognisable, otherwise it will not be acceptable. It should be divisible to allow sellers in turn to distribute their income between smaller transactions. Money should have a stable value to reassure those people who accept payment in money that they will be able to purchase a quantity of goods equivalent in value to the goods or services they have sold. Other characteristics of money include portability, for convenience; and durability, to enable individuals to save or delay making purchases. Durability and stability of value are important aspects of money for the building and property sectors, since these qualities enable money to be used to finance construction and housing and then spread the repayments over several years.

Money can take several forms. Indeed, sea shells, leather and cigarettes have served the function of money on occasion. Coin and notes are themselves only tokens. The need for liquidity in the economy calls for a variety of instruments. Cheques, credit cards and various bank and building society deposits all constitute liquid assets in different modern forms.

Attempts to control the money supply to bring down the rate of inflation led the government to introduce its Medium Term Financial Strategy (MTFS) in 1980. The MTFS sets monetary targets or ranges, giving an amount or quantity of money available in total in the economy. This is known as the money supply. However, as money takes several forms, the total supply of money depends on how money is defined. Currently two definitions of monetary aggregates are used – narrow money and broad money. Narrow money, defined as MO, is coin and notes in circulation and banks' own bank balances in the Bank of England. The main components of broad money (M4) are MO plus private sector bank deposits, and shares and deposits in building societies, less building societies' deposits in banks.

The subject matter of economics

In economics, everything stems from the implications of scarcity. Scarcity can be defined as the difference between our unlimited wants for such things as housing, clothing and food, and our limited means of satisfying these requirements. In economics, therefore, the problems of production, distribution and consumption are studied. Consequently, individuals acting within households, companies, governments or architectural practices are forced to make choices concerning what is produced, its quantity and its quality. Moreover, choices must be made to decide who should receive the goods and services.

The study of economics can be divided into microeconomics, mesoeconomics and macroeconomics. Microeconomics deals with the individual parts of the economy, especially the markets and marketplaces where buyers and sellers meet; the buyers may be seen as demanding consumers, while the sellers are the supplying firms.

Markets vary in size, geographical location and degree of competitiveness. They can range in size from small local street markets and shops, national exhibitions and trade fairs to large international money markets with thousands of participants throughout the world. Some markets are retail markets selling direct to the public, while others are wholesale, such as trade fairs, involving bulk buying and selling between manufacturers, wholesalers or retailers. In all markets, deals are struck. As a result of these deals, goods are delivered and services are provided.

Mesoeconomics deals with different sectors or parts of an economy and how they interact with each other. The economic problems, theories and issues of the construction industry are therefore discussed at the level of mesoeconomics.

Macroeconomics is the study of the economy as a whole. It deals not so much with individuals and particular products and services as such, nor with the specific problems facing an industry, but with the interaction of population, money and the total production and distribution of wealth. Macroeconomics is concerned with issues such as inflation, unemployment, international trade and government economic policy.

There is no such thing as a single economic solution for all societies, that could be applied to all countries. Economic systems depend on the political objectives of government and the interaction of political groupings. Economic systems may be seen to lie on a spectrum of systems ranging from centrally-planned or communist to free market or capitalist. In fact, most countries adopt economic systems that have elements of both types. To see this clearly, Figure 1.1 illustrates the spectrum of combinations showing that economies are invariably mixed, consisting of differing proportions of both public and private sectors.

The public sector in the United Kingdom is composed of central government, local government and public corporations owned and controlled by the government on behalf of the whole population, while the private sector consists of profit-orientated firms, from sole traders to public limited companies and charitable organisations. The government looks at public-sector spending in two ways in order to analyse and formulate policy. The first is to consider government spending in real terms. From 1950 to the mid-1970s the annual rate of increase in public-sector spending varied from just under 3 per cent to over 4 per cent per annum. Since the beginning of the 1980s, the rate of growth of real spending in the public sector has been less than 2 per cent per annum.

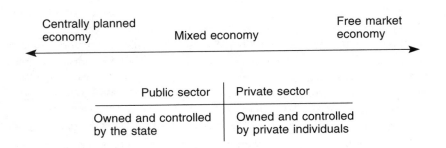

Figure 1.1   The range of economic systems

The second measure of public-sector spending used by the government is the percentage of the national income it consumes. Even if public-sector spending rises at the same rate as the economy as a whole, the percentage of national income spent by the government will decline. Since public spending is necessarily smaller than the whole economy, a proportionate rise in government spending will be smaller than the same proportionate rise in the national economy. Thus, even if real spending by the government were to rise, the increased spending would represent a smaller cut of the national 'cake'. In 1975, the public sector, including transfer payments such as unemployment benefits, grants and subsidies, accounted for 49.1 per cent of economic activity. By the mid-1980s, the public sector had declined relative to the private sector, to 45.8 per cent of the economy as a whole. This trend continued into the late 1980s. According to the Treasury (1994), the government planned to reduce its share of national output from 44 per cent in 1994 to 41 per cent by 1997.

In a centrally-planned economy all the land, the means of production (farms and factories) and the means of distribution (wholesale and retail outlets) are owned and controlled by the state on behalf of its citizens. All decisions concerning what is produced, its quantity, quality and price are determined by a centralised bureaucracy acting on the directives given to it by the government. As well as consumer goods such as food, clothes and domestic furniture, socially desirable goods may be produced regardless of the user's ability to pay, provided the government approves. For example, universal education and public health may be seen as desirable goods, which the state then funds.

The economy may be run according to five-year plans, designed to co-ordinate the production activities in all sectors of the economy, from agriculture to manufacturing. Targets and objectives are set by the government, against which actual performance can be compared. Because the market is not directly responsive to consumers' tastes and wishes, shortages of some commodities frequently arise, while overproduction of certain other goods may continue in order to meet government directives.

In recent years, several centrally-planned economies have undergone a radical change in their political direction, although several countries such as China, North Korea and Cuba remain predominantly state-controlled. Although few in number, these states still represent a third of the earth's population.

In contrast to central planning are free market systems, in which government intervention is kept to a minimum. In a free market economy everything is owned by private individuals acting on their own behalf. The profit motive provides the incentive for firms to produce goods and services to satisfy consumer demand. Prices are not determined by the government but by the interaction of many buyers and sellers in the marketplace. The distribution of goods in society is largely determined by individuals' ability to pay. The state does not intervene on behalf of the poorest. In free-market economies, while some individuals enjoy high standards of living, others live in poverty.

In economies with minimal state intervention, markets may often be dominated by single firms or groups of firms operating against the interests of the community by restricting their output and raising their prices. Where one firm controls a significant share of the market it is a monopoly, or where several firms combine, a cartel may be said to operate.

In fact, most countries lie between the two extremes of a purely state run economy and an unfettered capitalist system. Where they stand in terms of the spectrum illustrated in Figure 1.1 depends on the relative importance of the public and private sectors. Clearly, the People's Republic of China lies towards the left, while the USA lies towards the right. The UK is at present moving towards the right as a significant part of the public sector has been privatised and the proportion of the national income represented by the private sector increases.

Having described the broad framework of the economy as a whole, let us now turn our attention to the particular economic nature of the problems faced by building designers, building and quantity surveyors, and mechanical, electrical, structural and civil engineers. Later the role and economic contribution the building professions make to society as a whole will be discussed.

Economic constraints facing construction professionals

It is only within the last 150 to 200 years that an increasingly systematic approach has been adopted by building professionals towards building design, costing and management, following the start of the Industrial Revolution in Britain. Buildings encompass virtually every human activity. Building types therefore include not only houses, shops, offices and factories, stations, libraries, theatres, churches, museums and sports stadiums, but also civil engineering projects such as bridges, tunnels, roads and dams. These and many other types of building, standing alone or in combination, are the products of the construction industry.

Whatever the particular project may be, it is essential that it meets the constraints of economic viability, otherwise it will not be fully utilised and runs the risk of premature building obsolescence. Indeed, it is the architect's function to tackle with imagination the constraints of economic viability within the parameters set by the client, planning authorities, outside bodies and the architect him or herself. The economic constraints should not be seen as limiting factors but more as challenges and opportunities open to the architect/designer.

Stone (1988) noted that the designer occupied the central role. His or her contribution lay in meeting the needs of the client with a solution that was economic both to construct and to operate. Moreover, the architect may carry out his or her function with flair and imagination. It is this final, intangible aesthetic contribution of design, which, if recognised by the client, gives architects their unique position in the construction industry. The value of a subjectively attractive, functional and efficient design over a humdrum solution is the essence of the economic contribution of the architect. The best design is not always the cheapest.

It is for this reason that cost control has found its place in the building process, since the needs of clients to monitor costs throughout a scheme is necessary for completion of a project as specified, on time, and within a given budget. This is the role of the quantity surveyor.

Because of the complexity and the one-off nature of building projects, as well as the technical problems often encountered because of the varying ground conditions of every site, specialist structural engineers are needed to advise on methods of achieving a safe building of a particular design. In civil engineering projects, the civil engineer normally takes the lead role as the safety of the structure, rather than design, is of prime importance, although there are notable examples (especially of bridges) designed by architects, who have been supported by structural and civil engineers.

Increasingly during the 1980s, there has been a trend towards professionalising the management of the building process. This has been achieved increasingly by separating the role of management from the actual process through the use of management fee contracting. In return for merely administering the process, the management fee contractor receives a fee. The construction is then carried out by specialist contractors, often as many as forty on any given site, though not all at the same time.

The economic approach to design problem-solving

Economics provides one approach to problem-solving, based on comparing different options or solutions: to simplify the approach, assume that people behave rationally, and that they prefer to have greater rather than less satisfaction. We also assume that the objective of allocating scarce resources is to maximise expected usefulness or satisfaction – what economists call *utility*. How, then, is utility maximised?

It was noted above that a little more of one thing implies a little less of something else. When one chooses between such options, one is making decisions at the margin. A margin is an edge, and the marginal unit refers to the last unit of many. An extra unit of one good means one less unit of another. Because this concept of the margin appears several times in economics, it is vital to understand the term and its implications.

Total utility (TU) is the overall satisfaction derived from a product or service. Marginal utility (MU) is the change in total utility derived from the last unit consumed. A simple example will illustrate the concept of marginal utility.

In a domestic kitchen, for example, it will usually be necessary to install more than one electrical point. A second or even a third point will add a certain amount of convenience for those using the room. A fourth electric socket, although perhaps not as important as the second and third points, will add some more convenience. A fifth socket may not increase convenience at all, in which case, by the time the fourth socket has been installed, the convenience of having several sockets in the kitchen has been maximised. The difference each successive socket makes to the total convenience is the marginal utility of that socket.

The relationship between total and marginal utility can be seen in Figure 1.2. As the quantity consumed rises, total utility increases at a steady rate. However, the rate of increase gradually slows down until at Q_1, the total utility curve reaches a maximum at U_1 and one more unit of consumption adds no

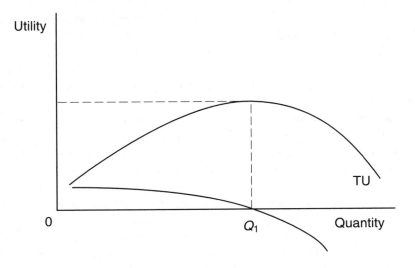

Figure 1.2   Total and marginal utility

extra satisfaction. Hence, the marginal utility of the Q_1 unit is zero and the MU curve is on the horizontal axis. Extra units of consumption beyond this point actually detract from total satisfaction, as further units lead to dissatisfaction at the margin. This is seen as the MU curve below the x-axis and a decline in total utility.

Figure 1.2 also illustrates the law of diminishing marginal utility, which states that (assuming everything else remains the same) as more units are used, eventually a point will be reached beyond which the increase in total utility derived from the last unit consumed will begin to decline. In the example above, this economic law applies after the third electric socket is installed. The question to ask is, therefore, what difference one more electric socket will make to total utility. Alternatively, one may ask, what difference would one less socket make to total utility. Provided that the marginal utility is positive, total utility will rise. Total utility falls when marginal utility is negative.

The price one is willing to pay is a measure of a product's marginal utility. The difference between what one would be willing to pay and the actual price is known as 'consumer surplus'. At the same time, it often happens that buyers would be willing to pay a price higher than that asked by sellers. Consumers will continue to buy a product as long as they receive a consumer surplus. Thus, if an individual is willing to pay £1.50 for a product (a pint of beer, say), but is only charged £1, the consumer surplus is worth 50 pence. As extra units of beer are purchased, the marginal utility of each successive pint will decline until marginal utility and therefore the willingness to pay, will be equal to £1. At that point the individual would still buy the drink, but it would be the last! Consumer surplus is, in a sense, a measure of the feeling of something *being worth it*, so often expressed in everyday conversation. This sense of value is also related to the term 'value for money', which will be examined in Chapter 7. Value for money takes the concept one step further by comparing the cost of alternatives to find the cheapest way of achieving a given end and therefore maximising consumer surplus.

Because there is a need to allocate resources between competing require- ments, there is a difference between maximising satisfaction from any one product and maximising satisfaction from the combination of all products purchased. Compromise is necessary. Maximising the utility derived from one product could be at the expense of satisfaction derived from other products, which in total could have been greater. To obtain the highest utility from a given budget, the more an item costs, the higher the extra utility derived from the last unit must be. The significance of the marginal utility concept is, therefore, that to maximise overall utility derived from all goods and services purchased, the marginal utility of each product is proportional to its price.

As long as one is willing to pay more than the price, extra units will be purchased. For example, if an individual consumer makes a subjective decision that the expected marginal utility to be gained from the consumption of one more unit of a good is comparatively high, then the price that individual would be prepared to pay for that extra unit would also be proportionately high. The main point here is that the decision to buy is based not on the *total* satisfaction, but on the extra satisfaction of consuming one more. If something is relatively cheap, then the consumer will continue to purchase units until the marginal utility derived from the last unit has declined in proportion to its price. Until that point, the ratio of marginal utility to price is greater than the same ratio for other products. The extra unit is 'worth the price' — one would be willing to pay a higher price than the price charged, because the extra satisfaction obtained from one more unit of that product is greater than if the money were spent on another product.

This approach summarises the application of economics to design problem-solving. For example, the varying wall finishes chosen for a given design will be chosen on this rational basis, if utility has been maximised. Thus the extra utility derived from the last unit consumed is directly proportional to its price:

$$\frac{\text{Marginal utility of last unit of rendering}}{\text{Price of unit of rendering}}$$
$$= \frac{\text{MU of last unit of brickwork}}{\text{Price of unit of brickwork}}$$
$$= \frac{\text{MU of last unit of concrete}}{\text{Price of unit concrete}}$$
$$= \frac{\text{MU of last unit of any product}}{\text{Price of any product}}$$

If any product is out of proportion, the rational consumer will either increase his or her purchases of that product to reduce the marginal utility and return it to the same proportion as other goods and services, or abstain until the marginal utility to be derived from one more unit is again in a similar ratio to price as the alternatives that s/he might have purchased.

The decision is not just to purchase, say, tomatoes, oranges and potatoes, but to decide the quantities of each to be bought. The quantity of tomatoes therefore depends on the price. If marginal utility diminishes with each successive tomato added to the bag, the consumer will stop filling the bag at the point when the expected marginal utility of the last tomato is exactly equal to its price. The rational consumer would then go on to do the same with the oranges and potatoes.

Where the expected marginal utility is low, and the price relatively high, only a few units may be purchased by the consumer. The reason for this is that it would be possible to gain more extra satisfaction by spending the same money on one more unit of an alternative good or service, which would provide higher marginal utility.

Although difficult to master, the concept of marginality is one of the most important concepts in economics. The concept of the marginal unit occurs not only in conjunction with extra satisfaction, but also in costing, productivity and in welfare economics.

The nature of choice in the construction industry

The problem of scarcity occurs in construction projects, namely, the problem of having insufficient funds to do all the things one would like. Compromise is necessary and decisions have to be taken. Certain design elements may have to be omitted. This is the constant predicament in which clients and architects find themselves, and this pressure compels the responsible architect to produce cost-effective designs.

Architects may be faced with similar decisions: for example, they may be faced with a choice between expanding into a second office or buying extra equipment. Funds may not extend to doing both immediately. Each option has its advantages and disadvantages and in choosing one, the architect has given up the opportunities offered by the alternatives. Hence, the true cost of a chosen option is the next best option and its net benefits that have been sacrificed. This cost is therefore called the *opportunity cost*. Opportunity costs will be discussed in more detail in Chapter 7.

Just as the aims set by the client determine a building's most economic shape, the aims of the individual partner in the practice will ultimately determine the most economic combination of work and people for that office. In other words, there is no one way to run a practice, because the most economic method depends to a large extent on what those in the practice may be hoping to achieve.

This chapter has illustrated some of the economic linkages between the construction professionals and the construction industry and the construction industry, and the economy as a whole. Not only are events in the economy of great importance to the industry, but the interaction of the industry with the rest of the economy has occasionally tempted governments to use the building industry to stimulate employment.

Moreover, the building industry reflects levels of demand in society and the economy. An appreciation of these will aid informed estimates of future trends. External influences on the building industry include the distribution of wealth and the development of new industrial products and techniques, which may stimulate the construction of factories. The attitude of government to state intervention in the economy and state responsibility for welfare will have a major impact on demand. The growth of the economy and expectations of industrialists, bankers and consumers concerning the future also have to be considered. This is the subject matter of Chapter 3, but before we look at the current situation we shall take a look at the developments in the construction industry that help us to understand how we arrived at the present position.

Part 1
The Construction Sector

2 The Construction Industry in a Historical Context

Introduction

Many work practices and relationships in construction at the end of the twentieth century appear to be haphazard, illogical and even downright inefficient. To understand the relationships between firms in construction and the organisation of work in the building industry it is best to look at how these relationships and methods arose in the first place, and how they developed into the current situation.

Construction before 1800 – an account of the guild system

During the medieval period, between the approximate dates of 1200 to 1600, construction work was undertaken by master craftsmen organised into guilds. Each guild represented a different craft, and each craft was based on a different material such as lead, wood or stone. In this way craftsmen were divided into trades.

The guilds controlled the prices that master craftsmen could charge for finished work, so the craft guilds sold the products of their members rather than their labour. The guilds not only controlled 'fair' prices, which took labour time into account, but also set standards and regulated quality. However, stonemasons, who were organised into lodges and were mobile, were the exception: they charged for their labour.

The guilds dictated how many journeymen or skilled workers a master could employ, and set the wages they were paid. Apprentices paid for their training, which usually lasted for seven years. This system of entering a trade ensured that only those from wealthier families could afford to become master craftsmen and uphold the status of the guild. Moreover, to become a master craftsman, it was necessary to be a citizen of the town. In this way the guilds prevented outsiders from competing, unless their skills were required. Competition between the masters themselves was based on quality and reputation rather than price. They were working proprietors who sold their products or services to merchants. Finally, master craftsmen also frequently employed labourers. This was the general pattern of production of goods and services under the guild system.

For the construction of large projects, a Clerk of Workers or a Master of Works or a Keeper of Works was appointed. The purpose of the Clerk of Workers was to oversee the whole building process on behalf of the client. The structural material used determined which master craftsman became the dominant master and the dominant craftsman on a building usually became the chief master. Often this meant that the stonemason became the chief master mason, who in turn oversaw the other master craftsmen. The masters remained separate as each master craftsman was jealous of his own material.

On smaller buildings no one was in overall charge of construction. For the design, employers or clients would simply refer to other existing buildings and continue to build a little at a time as long as their money lasted, or until the building was suitable. Early vernacular buildings were made from wattle and daub, which was a mixture of mud, horse manure and straw. Bricks were rarely used before 1550. Building materials were taken freely from common land.

From around 1600, as materials became relatively scarce and harder to find, they began to be sold to the master craftsmen by merchants. Over the next two hundred years 'master builders' emerged as a broader term to describe chief master craftsmen who took on increased responsibilities for the co-ordination of the building process. The master builders were, however, still master craftsmen not contractors. Each craftsman had a separate contract, although increasingly these contracts were based on day rates, established by custom and tradition.

The client still designed the building as construction progressed, although, following the Great Fire of London in 1666, various building acts were passed by Parliament to prevent a repeat disaster and to improve building standards. The Building Act of 1774, for example, set out the 'proper thickness of Party-walls, External-walls and Chimneys'. However, prior design, pricing or measuring were still uncommon.

Construction during the 1800s

From the late eighteenth century and in the early nineteenth century, merchants began to buy land, and a market for land developed. The price of construction work became an issue increasingly outside the control of the guilds. Around 1800, the contracting system emerged and many contracts required prior design for materials, labour and costs to be estimated in advance of actual building work. This was especially true of speculative building since the risk of not finding a buyer after completion of a project would mean losses to the builder. To find a buyer and sell at a profit, costs had to be controlled.

Speculative building in Georgian London followed a pattern based on leasing land. A lease is a contract to rent land for a given period, often 100 years. Planning was carried out by the landowner, while merchants who leased the land placed subleases on plots with builders, who had to bid for the Contract in Gross. The Contract in Gross was based on pre-design, pre-measurement and competitive bidding, to find the market price for the contract. On winning a Contract in Gross, a master carpenter, for example, could take on the financial responsibility and risks of building small numbers of houses by subcontracting the building work to master builders. The master carpenter acted as the main contractor in this instance, and his profits came from the bid cost of construction. Although the main contractors used sub-contractors, they also employed their own skilled workers and navvies or unskilled labour, all at market wages.

Foremen were used instead of master craftsmen. Internal subcontractors were paid a 'lump sum' to find labour – a system used in other industries, including mining and textiles. The artisan system of masters, journeymen and apprentices was replaced by a system of employers and labourers. One of the largest employers to emerge at that time was Thomas Cubitt, whose building firm survived well into the second half of the twentieth century. Similar methods of working were used in civil engineering, especially in railway building.

As the old guild system had been in decline for many years, in 1814 an Act of Parliament abolished the medieval system of apprenticeship. Training was left to firms themselves, who set up their own schemes for new entrants to construction, who then became skilled labourers. Building manuals were published, including *The General Principles of Architecture, of the Practice of Building, and of the several Mechanical Arts connected therewith,* by Peter Nicholson, a builder and mathematician. A sixth edition of Nicholson's manual was revised and corrected by Joseph Gwilt in 1848.

The building sector was responding to demands placed upon it by the Industrial Revolution, which had begun to change the nature of clients. The new and growing markets, inventions and political circumstances gave people the opportunity to profit from the changes. Many buildings were now built on a speculative basis, to be sold for profit on completion. Often new buildings were part of a business plan, such as mills and factories, so that both price and speed of construction became important.

The emergence of the architecture and surveying professions

These changes in the market for buildings and construction services led to changes in the way the building process was organised. The profession of the architect, quite separate from the building trade, has existed since about 1835 when the Institute of Architects was established. This professionalisation of architects was in response to unscrupulous individuals who, for example, would modify designs and claim the credit for the completed buildings. Both clients and architects needed the protection of a professional body and a code of practice, part of which prohibited Fellows of the architecture institutes in both England and Scotland from having any financial interests in the building trade.

Until then, in Scotland, architects had been employed primarily as builder/ designers. Very few Scottish architects, apart from Robert and James Adam, James Playfair and one or two others, had been able to carry on a livelihood solely as architects. They had gone to London in search of the larger market. Indeed, the architect was hardly a respected member of society, coming well down the social hierarchy in Edinburgh society in the eighteenth century. The fees architects attempted to charge were around 5 per cent of the total building cost, based on 1 per cent for design and 4 per cent for supervising the construction.

Since the seventeenth century, wrights or masons had been appointed as sworn measurers by the City authorities in Edinburgh. Colvin (1986) describes their role as impartial measurers of building work, an early form of public arbitration in the building industry. In England, a person employed on a building site was often paid by measure rather than by the day. They were paid according to what they produced rather than how long it took them to produce it. Either they would have to accept their mason's or carpenter's calculation of their own work, or they could, at their own expense, engage an independent surveyor. In Edinburgh, however, it was possible to call on the services of the sworn measurer, who was a public official, remunerated at a fixed rate by the City Council.

During the Napoleonic Wars towards the end of the eighteenth century, measurers allowed 15 per cent for profit. As the nineteenth century progressed, contractors had begun to compete for the job of constructing the whole building. This enabled architects to concentrate their efforts on design and representing the client's interests. Architects, according to Powell (1982), were increasingly employed on a fee basis to provide detailed drawings, specifications and tender documents to enable contractors to compete for work on a project. In 1919, these fees were 6 per cent of the cost of works. The measurers of old were replaced by the new role of quantity surveyors, appointed by the contractors. Usually, two surveyors were appointed, one to look after the interests of the contractor and the other to represent the client.

By the 1830s the introduction of contracting and the separation of professional roles had led to the establishment of the Federation of Master Builders, the Institute of Chartered Surveyors, and the Royal Institute of British Architects. Although trade unions had been banned by the Combination Acts even before 1800, friendly societies were formed to help their members in times of need. In 1824, the Combination Acts were repealed and eight years later the first national building operatives union was formed. However, it was not until the very end of the century that employers organised themselves on a national basis. The National Federation of Building Trade Employers was formed in 1899. This paved the way for national negotiations and in 1920 the National Wages and Conditions Council was established, which became the National Joint Council for the Building Industry in 1925, with meetings between employers and unions at national, regional and local levels.

Construction during the 1900s

In the mid-1960s to early 1970s there were approximately 3600 architectural practices in the United Kingdom and 700 practices of quantity surveyors. These figures contrast with the 70 000 building firms in Great Britain in 1973, though the vast majority of these were very small firms. In 1991, the number of building firms had risen to 207 000, though the vast majority of these, 174 000 firms, comprised only one to three people. Fewer than 100 private contractors employed more than 600 people in 1991.

In the post-Second World War period surveyors on the design side tended mainly to be members of the Royal Institution of Chartered Surveyors (RICS), while those on the building and construction side mainly belonged to the Institute of Quantity Surveyors. In 1983 the Institute of Quantity Surveyors merged with the RICS. Surveyors' functions have now developed and diversified. Quantity surveyors prepare bills of quantities to enable contractors to tender and they also monitor progress by pricing and measuring work carried out, and settle final accounts. Today they are increasingly involved in cost planning during the design stages to ensure value-for-money design solutions, and perform many other roles, including advising on building procurement and contractual matters, and settling building disputes. They also play a major role in project management.

Changes in the construction industry labour force

It is revealing to note the changes that have taken place in the labour force of the construction industry since the beginning of the twentieth century. The number of people employed in the construction industry has fluctuated from year to year. In 1948, 1.45m were employed rising to a peak of 1.8m in 1968. By 1973 the numbers had declined to 1.6m, of which approximately 55 per cent were employed by private contractors, 15 per cent were employed directly in the public sector, just over 7 per cent were self-employed, and 23 per cent were involved with administration and clerical work.

In 1914, the woodworkers had by far the largest union, with 79 000 members. After the First World War, the reorganised Amalgamated Society of Woodworkers continued to grow and by the mid-1960s its membership stood at 122 000. In 1973, 13 per cent of all employed operatives were carpenters and joiners, still by far the largest craft, followed by painters with 9 per cent, and bricklayers with 7 per cent. Electricians, plumbers and many other trades made up the remainder, but not in such large numbers.

Over the long run, material costs have tended to rise less rapidly than labour rates, though material prices do occasionally rise faster than wages, as occurred in 1994. By 1979, according to the Architects' Journal of 5 September of that year, the guaranteed minimum weekly earnings were £67.20 for craft operatives in London and Liverpool, and £67 in Scotland. Labourers could earn £57.40 in London and £57.20 in Scotland. In 1991, according to the Housing and Construction Statistics, male full-time manual construction workers (including both craft operatives and labourers) earned on average £257.10 per week, including overtime, compared to £253.10 on average for all industries, in return for 45.4 hours of work per week in construction compared to 44.4 hours on average in all industries. Although there was a real increase in take home pay, this rise in wages also reflects changes caused by inflation. Nevertheless, excluding the effect of overtime on pay, construction workers earned slightly less on average than workers in other industries: £5.51 per hour compared to £5.54.

In the early 1990s the difference between the pay and hours of work of non-manual workers in construction compared to other industries is more marked. In return for working an average week of 40 hours compared to 38.7 hours in other industries, non-manual workers in construction earned £368.20, against the industrial average of £375.70. Construction workers earn less in return for longer hours than the average for all industries. This does not necessarily imply that construction incomes and conditions are the worst, only that they are below average. Figures for female remuneration and hours are not given in the Housing and Construction Statistics.

The construction industry must compete for labour with other industries, which have been able to increase the wage rates they offer because of rising productivity in factories, partly related to the introduction of mass-production techniques in the nineteenth and early twentieth centuries. These techniques are sometimes referred to as 'Fordist', after Henry Ford's introduction of the assembly line in car production. Even today, the introduction of computer technology and robotics continues to raise industrial labour productivity. This may allow incomes to rise while reducing costs per unit produced. In this way, manufacturers are able to produce new materials and goods at reduced prices.

Demand for construction in the 1900s

Just as house building reflects population movements, industrial building is carried out in response to increased demand and output of industry. From the beginning of the 1920s, house prices rose and the volume of housing output grew. Owner-occupation expanded with the increasing significance of building societies, the availability of cheap credit, and mortgage subsidies. Subsidised public-sector housing for rent also increased. In the 1920s developers bought land, built houses and sold them. In the following decade several building companies such as Costain, Laing and Wimpey grew rapidly, and in the 1930s established themselves as very large concerns as a result of winning major military contracts such as airfields and harbours.

Since the Second World War, when 58m sq. ft of commercial property was demolished and the number of houses bombed reached 475 000, there have been several phases of heightened building activity in the UK. From 1939 to 1954, public-sector demand dominated. After the war, the government exercised rigid control over building, giving priority to war damage repair, schools and local authority housing but a balance of payments crisis in 1947 led to a cut in public-sector house building of 50 per cent. In the early 1950s building restrictions were relaxed to some extent, though the location of buildings was still controlled. Building permits were required by the private sector, and the Development Land Tax was introduced to tax developers' profits. However, after 1954, restrictions were removed and in the 1960s office speculation began and private sector housing expanded. In the 1960s and 1970s many urban fringe, high-density housing estates were built. Older houses were also split into flat conversions. In the 1980s fewer fringe sites were made available, but the prices of up-market properties rose.

From 1955 until the early 1970s, the workload of the construction sector varied by about the same amount as the average for other industries. From the late 1960s to the early 1970s there was a property boom, which came to an end with the oil crisis of 1973. From the mid-1970s and throughout the 1980s, construction demand was far less stable than it had been in the previous decades. In 1973–5, demand dropped by 30 per cent, and in those years construction output probably varied more than in other industries in terms of price if not in terms of volume of work. In 1981, and again in 1990, there were similar reductions in construction demand. Although construction firms relied on government spending for expansion, in 1975 and 1976 economic crises led to cuts in public spending on housing and civil engineering. This led to a downward shift in public-sector spending as a proportion of all spending in the economy, a trend which continued throughout the 1980s. Activity in the building industry is inextricably linked to the rest of the economy and to government policies.

Just as the Simon Report had stated in 1944, at the beginning of the 1980s it was again argued that the economy and employment could be stimulated through the construction industry, by creating building projects. These projects could have been financed through government borrowing, an economic strategy adopting Keynesian policies. John Maynard Keynes (1883–1946), who had been the leading British economist in the first half of the twentieth century, had argued that in recessions governments need to spend more than they take in through taxes in order to create sufficient demand in the economy to reduce unemployment. This policy of a government deliberately spending more than its tax revenues is known as *deficit budgeting*. However, in the 1980s, there was no Keynes available to convince the government to pursue a budget deficit policy, and the government refused.

Although for most of the 1980s the government did in fact end up with a budget deficit caused by high levels of unemployment, it refused to stimulate the economy by increasing its spending on construction projects as a whole, although several motorway projects were initiated in this period. In any case, in the 1980s, stimulating the economy by investing in construction projects may not have been very effective. During the 1980s the construction sector came to rely increasingly on using building components from abroad. As a result, any attempt to stimulate construction activity would have drawn in a large volume of imports. In the 1980s, the government stimulated the economy at various times with tax cuts rather than construction investment.

The boom in construction in the late 1980s peaked in January 1989, though construction work started before then meant that workloads continued at a high level for several months. Output in 1989 was similar to 1979, though building costs and prices were much higher. Construction demand was mainly for shopping malls and office construction, especially in London and the South East, although other provincial cities followed suit. Annual housing output was still lower than in 1969 or 1972.

Construction industry cycles

Having described the fluctuations in the construction industry since 1945, certain patterns emerge, and a theoretical framework may help to provide an understanding of some of the causes of these variations in demand for construction work, although a proportion of the variation still appears to be caused by unpredictable events outside the scope of an economic theory. The trade cycle nevertheless describes the variation in levels of economic activity over time. Trade increases or decreases. It rarely stays the same from one year to the next. Figure 2.1 illustrates the trade cycle schematically, showing the sequence of recessions followed by periods of recovery. The actual level of activity in the economy may be measured using employment statistics, as employment increases during recovery periods and declines during recessions.

Economists have long been engaged in analysing the duration and pattern of trade cycles to predict levels of activity and demand in the economy. For example, Simon Kuznets identified building trade cycles which lasted between 15–25 years from one boom to the next. Others have argued that the housing trade cycle follows a bulge in the birth rate: for example, after a war. When these babies grow up and leave home approximately 18–20 years later, they stimulate demand for housing. In turn, they too have children. This bulge in the population statistics behaves like a wave, stimulating demand every twenty years or so, as each generation leaves home to find new accommodation.

The supply of credit and the rate of interest are especially important in the construction industry, which relies on easy access to funding for long-term projects. Changes in the size of inventories and the stocks of goods held by companies in reserve also affect the demand for new building. These factors will be examined in more detail later on.

The construction industry is particularly vulnerable to fluctuations of

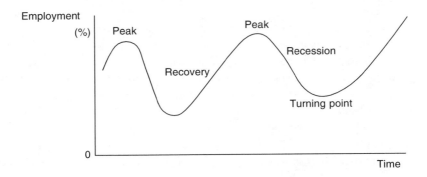

Figure 2.1 The trade cycle

activity. The problem is that the output of the construction industry is durable. Buildings last for many years, and therefore decisions to place orders can often be delayed. When firms are in financial difficulty, one of the things they will look at is their investment programmes to see if there are any projects that can be postponed or even cancelled. Also, because buildings are durable, they continue to stand even after they are no longer required. The number of empty buildings is a measure of the spare capacity of the existing stock of buildings at any one time. During the early part of the 1990s, vacant office space in London not only depressed rental values and property prices but also reduced demand for construction services. These were not required until developers could once again see opportunities for renting or selling new or refurbished buildings.

It has been argued that the building trade cycle goes through various phases. A residential building phase was followed by increased profits for manufacturers and retailers, which in turn led to a boom in office and shop building in the later part of the building cycle. Spontaneous, irregular, contingent events may also cause construction activity to increase or decline from one year to the next. Population changes stimulate building booms, as, for example, when people move from agricultural employment in the countryside to seek employment in manufacturing industry in towns. Other demographic factors which affect demand for buildings include the population's age structure, marriage rates, and household formation. Household formation is related to the number of people who leave home annually or seek divorces, or separate. These social trends have consequences for the demand for accommodation, as do changes in the level of unemployment, the distribution of income, interest rates and house prices. The application of new technology may act to stimulate activity in specific sectors of industry and commerce, which in turn creates work in other sectors of the economy, such as the building industry, while decline in other industries can have the opposite effect.

The workload of construction professionals is far from steady, because it is subject to variations in demand caused by the trade cycle in the rest of the economy. Periods of growth and new commissions are followed by periods of recession when clients are harder to find. In recent times, the most difficult periods for architects, for example, have been between 1974 and 1977, and from 1989 to 1994. From 1973, new commissions and the number of production drawings declined steeply. Architects had expanded their offices to cope with demand up to 1973, only to find that they could not maintain their offices and capacity to carry out work because there were insufficient orders for new work. In 1977 the industry began a period of recovery which lasted, apart from the downturn in 1980, until the late 1980s.

Several theories have been suggested to account for the upturns and downturns in the level of economic activity. Of particular relevance to the construction sector is the theory of the accelerator principle. Demand for consumer goods, such as furniture, shoes and radios, varies from year to year. For example, demand may increase steadily for a period at, say, 3 per cent per annum, which is the rate of change in demand. The accelerator principle demonstrates how a small change in the rate of change in demand for consumer goods leads to large changes in demand for investment goods, such as factories and shops, for example when the rate of increase in demand slows down from 3 per cent per annum to 2 per cent.

The concept at the heart of the accelerator principle is the capital–output ratio. Capital is the amount of plant and machinery, money and materials. Output is the value of goods or services produced in a year.

$$\text{The capital–output ratio} = \frac{\text{Capital}}{\text{Annual output}}$$

This ratio shows the amount of capital required for a given quantity of output. If the capital output ratio is 3 to 1, for every £100 000 worth of annual output, £300 000 of plant and machinery is required. If output increases by £5000, then the firm would need to invest £15 000 to maintain the same capital to output ratio. This can be seen in Table 2.1 below.

Each year firms need to invest in plant and repairs simply to maintain output at its previous level. They also need to invest in additional plant and machinery to increase their capacity to meet increased consumer demand. Assuming shops, offices and factories last on average thirty years, then approximately 3 per cent of these buildings will need to be replaced each year. Otherwise, after thirty years, there would be no usable space left. If consumer demand rises by 3 per cent, then total construction demand would rise to approximately equivalent to 6 per cent of existing building stock (3 per cent for maintenance and repair, and 3 per cent for the extra capacity), and this level of construction demand would last as long as the demand in the rest of the economy continued to expand at 3 per cent per annum. If, however, demand in the rest of the economy rose by only 2 per cent instead of 3 per cent, after having risen at 3 per cent for several years, although consumer

Table 2.1   *Capital–output ratio*

Capital–output ratio	Capital required (£)	Annual output (£)
3:1	300 000	100 000
3:1	315 000	105 000

demand would still be increasing, demand for construction would drop to the equivalent of approximately 5 per cent of existing stock, because the need for extra capacity was less than that previously required. This would lead to a drop in construction demand of approximately one-sixth, down from 6 per cent of existing stock, in spite of a rise in consumer demand of 2 per cent. This drop in demand for investment goods such as buildings and machinery can set in motion a downward trend in the whole economy, as firms in the rest of the economy see construction and investment demand in decline.

Although this is an oversimplification of the economy, it indicates the nature of the relationship between construction and the rest of the economy. In fact, the accelerator principle can be used to explain the ever-changing nature of an economy such as that in the UK. The following is a stylised version of events showing the relationship between the construction industry and the rest of the economy.

Even when the economy is growing and employment is relatively high, demand for buildings can slow down if firms feel that the rate of economic expansion cannot be sustained. This leads to a drop in the demand for building work, especially new build, which in turn results in unemployment among building workers. As firms outside the construction industry observe a slowdown in construction, they become cautious in case they overproduce and cannot sell their outputs. This causes a further slowing down in the economy, and increases unemployment. Demand declines as unemployment rises and this decline makes firms reduce their demand, not only for extra buildings but also for repair and maintenance of many of their existing buildings, which may be surplus to requirements.

However, when a decline in the economy as a whole has been set in motion, the decline does not last for ever. Eventually the rate of decline itself slows down, even though the economy may still be declining and unemployment still rising. The proportion of the population out of work cannot rise for ever, or the whole population would become unemployed. Once the rate of decline slows down, firms need to increase their repair and maintenance, or the number of existing usable buildings would continue to drop. This increase in construction work occurs even when the economy is still declining. Construction workers are taken on. Unemployment in construction may decline. Demand from construction workers for clothes and furniture, holidays and food increases, and firms in the rest of the economy begin to benefit. Eventually, the firms in the rest of the economy regain their confidence in their respective markets and decide to invest in new buildings and increased capacity. This process now sets the economy back on a path of expansion and the trade cycle begins again.

Developments in materials, technology and methods of construction

Bowley (1966) pointed out that the traditional procurement method which used the architect as the project leader meant that the design of a building was carried out before the contractor was appointed. The fees of the architect were based on a percentage of the contract price. Until the contractors received the tender documents, they could not price the work; they did not know the size, type of structure, design or materials and therefore could not predict the method of building or the skills and machinery needed. This made it difficult for contractors to plan or invest before their next project was known.

There was therefore a separation of designers and builders. While the design team worked in professional architectural practices, the builders who put the designs into practice worked in contracting firms. Although architects had the opportunity to innovate and reduce costs, they had little incentive to do so, because they were paid a percentage fee based on the cost of construction. The builders, on the other hand, who had an incentive to invest in new equipment in order to undercut their competitors prices, had little opportunity to innovate, because they had to adhere to the designs and specifications laid down before the tendering stage.

Bowley therefore saw the traditional method of procuring buildings as an obstacle to innovation. Innovation is the practical application of inventions in order to reduce costs. However, in construction, unlike other industries, reductions in costs are not usually passed on to final building owners or users. A reduction in the expected cost of construction would allow a developer to raise the bid for a given site. As all potential developers would benefit from reductions in construction costs, they need to raise their offers in the market for land. If land prices rise and the final cost of a finished building including its site remains unaffected by the lower cost of construction, then there is no financial gain to the building owner or user. It is other factors, such as the state of the economy, which cause building prices to vary.

Contractors also face a disincentive to innovate. Ball (1988) has argued that as each project is a one-off, it is not possible to predict a firm's next project and its method of construction. As innovation involves research and development, and this in turn involves expenditure and risk, it is always highly uncertain if a contractor would be able to apply any radical new ideas to his/her next project. However, package deal firms often provide standard products, which clients can take or leave, and to some extent these can incorporate innovative ideas. Also, in design and build contracts, contractors take responsibility for both the design and construction. However, design and build increases the builder's professional liability and this slows down innovation in order to reduce risk. Buildings such as supermarkets, warehouses and advance factories are usually kept simple, often steel frame, and quick to erect.

There are, however, many examples of innovation in the specialist contracting sectors, component and equipment manufacturers, and materials. For example, piling and drilling techniques, road laying and steel erection, plate glass fixing, and cladding techniques are in a constant state of change and development. Plant manufacturers are constantly introducing new models and as new replacement equipment is introduced it invariably embodies the latest improvements, thus reducing energy, maintenance or labour costs.

Innovation in the use of new materials, components and methods of building therefore shifted to manufacturers and suppliers to the construction sector, although the public sector has often led innovation in construction, seen from the client's point of view. For example, various building systems were produced during the 1950s, including a lightweight steel frame system called CLASP, which was developed and built by a consortium of local authorities. A building system is a standard method of construction capable of being applied to different building sites. Some systems relate to components, while others are concerned with the whole structure. Seeley (1996) has described a number of systems, ranging from load-bearing cross wall construction using concrete, bricks or concrete blocks to composite systems with steel or concrete frames. The purpose of system building techniques is to standardise methods in order to reduce building costs and the risk of building failure, and to increase the degree of prefabrication in order to speed up the construction process.

Innovation in construction, such as system building, has often been the result of a shortage of materials, or the introduction and application of new materials, or the need to meet a particular demand for buildings. As prefabrication of building components has increased, so have the number of outside firms supplying their own specialist teams and plant to supply and fix their own products on site. In this way new technology and techniques permit more and more construction work to be carried out off-site, in the factories of manufacturers. The main activities of on-site labour consist of preparing the site, handling materials, and assembling and fitting components. As much as 60 per cent of the value of construction work is carried out off-site by suppliers. Apart from the traditional inputs of timber, bricks and cement, the steel, plastics and glass industries also act as suppliers to construction. Competition between firms in those industries, and between industries producing substitutes for more expensive materials or components, has led manufacturers to be innovative.

According to official indices published in Housing and Construction Statistics in 1989, in the period from 1978 to 1988, all material prices rose, though some prices rose faster than others. For example, the cost of bricks to builders rose by 212.2 per cent; the prices of products and materials as diverse as uncoated sand and gravel aggregates, copper tubes, and thermal and acoustic insulating materials rose by around 169 per cent; the price of wooden window frames rose by 145.8 per cent; and the price of metal doors and windows rose by 132.7 per cent. Only the price of certain concrete pipes rose by as little as 12.4 per cent in the period. The method of construction, and the structural systems and materials selected for building, all depend on prices. As prices vary relative to each other, alternative methods of building are used to take advantage of cost differences. An efficient method of building in one period may not be efficient in another, especially when the cost of materials of one method becomes more expensive than the cost of materials of an alternative.

In house building, for example, a standard product is often developed by builders. Its appearance may be traditional, although new methods are gradually introduced, often copied from the public sector. Timber-frame houses were used in the public sector before speculative house builders adopted the method in the 1970s and 1980s. Other innovations include prefabricated roof trusses, windows, trims and finishes. Manufacturers of steel and glass design and fit curtain walling, bolting on large prefabricated sections, reducing site labour and speeding up the site construction process.

Standard components have increasingly been adopted because prefabrication implies that when components are eventually brought to site they can easily be assembled. However, it is not sufficient to have standard components; it is also necessary to have common interfaces between these different components. Components need to be not only easy to fit, but easy to fit correctly, even with components from different manufacturers. This highlights the need for modularisation. Many components from manufacturers may work alone but fail when part of a system. Specialist trades and subcontractors also innovate but, again, buildings may fail at joints, where two trades meet.

Offshore steel and concrete technology has already begun to use large, prefabricated components or modules. Modularisation implies that there is a standard interface between components and that they are easy to fit. In the Hong Kong Airport project, concourse units measuring as much as 72 m × 36 m × 36 m and weighing 4500 tonnes were produced off-site. Other modules included lifts and escalators. The benefits of modularisation for the contractor include significant reductions in congestion on site, and site labour, with reductions in labour transportation costs of getting people to the site, as well as reductions in accommodation and subsistence costs. Modularisation also simplifies maintenance as building components can be inserted or replaced, like pre-wired circuit boards. In this way, modularisation is deskilling traditional crafts, although it might be argued that it calls for new skills in the design and installation processes.

Machinery in construction is often developed from borrowed technology used in other industries, such as pit props in mining. However, earth-moving equipment is one class of machine that was developed specifically to meet the needs of contractors. In civil engineering machines can be used to replace labour more easily than in building because of the repetitive, continuous nature of many civil engineering tasks, such as road laying and tunnelling. Also, compared to building, civil engineering projects often involve fewer diverse operations and lend themselves more readily to the greater use of mechanical plant.

In the UK construction industry general plant and machinery is usually hired. The more specialist the firm, the more it tends to own its dedicated equipment. Hiring plant reduces the need for an outlay of capital and helps contractors to ease their cash flows, charging the hire of plant to each project. Many plant hirers are, in fact, subsidiaries of contractors, hiring out plant when it is not required by the parent contractor.

The growth of plant hire in the UK is the result of the need of contractors to be increasingly flexible in their ability to respond to changing market and site conditions. In the 1980s the increase in subcontracting and the introduction of new procurement methods, such as construction management and management fee contracting, forced contracting firms to reduce their directly employed labour and directly owned plant.

Because of the increase in prefabrication, a far smaller proportion of the value of construction output is carried out on site. This means that statistics have shown a decline in manpower in construction. However, this decline reflects shifts in work from site to factory and the increasing use of mechanisation on site as well as simplification of the assembly process. This increasing use of prefabricated components reflects the shortage of skilled labour and therefore its increased cost.

Professional, trade and public-sector organisations in construction

House's Guide to the Construction Industry provides over 1000 organisations linked to the production of the built environment. Professional and trade organisations in construction can be grouped into associations of professionals, trades and unions.

Professional bodies involved in design, planning and management of construction projects include the Royal Institute of British Architects, the Royal

Institution of Chartered Surveyors, the Architects and Surveyors Institute, the Chartered Institute of Building, the Incorporated Association of Architects and Surveyors, the Society of Architectural and Associated Technicians, the Society of Surveying Technicians, the Institution of Civil Engineers, and the Institute of Clerks of Works.

These institutions and societies aim to promote a professional image of their members to the general public. They give professional recognition to members, who must pass examinations to join. They lay down standards of professional conduct. They attempt to ensure that price competition between their members is kept to a minimum, and assist their members in marketing their services by providing a first port of call for potential customers who may be given advice and contacts. In this respect they are very similar in form and practice to the traditional craft guilds. They are now becoming increasingly market-led, responding to changes rather than imposing sets of rules which they are unwilling to alter.

Trade organisations include the Building Employers' Confederation, the Federation of Master Builders, the Specialist Engineering Contractors Group, and the Construction Plant Hire Association. Craft associations include the British Woodworking Federation and the National Federation of Painting and Decorating Contractors.

Federations of employers and trade associations represent the joint interests of their members. Their function is to present, defend and promote these interests by publicity information and representations to Parliament as well as organising and informing their members when co-ordinated action is required. They also defend the position of their members in national negotiations and when proposed legislation is likely to affect their membership. By protecting the interests of their members, professional bodies are to firms and professionals what unions are to the workforce. It is far more effective if a representative body is in a position to call on the support of its members than if each member is left to fight alone.

The main *trade unions* include the Union of Construction, Allied Trades and Technicians (UCATT), with over 100 000 members in 1996, representing brick-layers, painters, stonemasons, joiners and other crafts. The Transport and General Workers Union (TGWU), with approximately 1.9m members in total, has a section for construction workers, including plant operators, scaffolders and roofers. The General and Municipal Workers and Boilermakers and Allied Trades Union (GMBATU) also have members, mainly non-craft operatives, who work in the building industry, especially in local authority direct labour organisations. Other unions with membership in the construction industry include the Amalgamated Union of Asphalt Workers, and the Furniture, Timber and Allied Trades Union (FTAT). Finally, electricians and plumbers are represented by the Electrical, Electronic, Telecommunications and Plumbing Union (EETPU).

Since the early 1980s, the trade union movement in the construction industry has been very weak. This weakness has arisen for several reasons. The growth of casual self-employed labour has meant that many workers have been excluded from membership of a union. Labour legislation in the 1980s further weakened all unions and made it increasingly difficult to organise workers on construction sites. The 1980 Employment Act provided funding for secret ballots. Sympathy strikes and secondary picketing were made illegal in the Trades Unions Acts of 1982 and 1984 respectively.

Construction sites are geographically dispersed, temporary and employ a fluid workforce. Moreover, many building sites are small, employing very few people. These factors make it difficult for unions to find new members and then communicate with them. Moreover, many employers and a proportion of the workforce itself is at best indifferent and at worst hostile towards trade unions. The employers see unions as a threat to industrial peace on site, because hazardous working conditions, low pay and job insecurity would be challenged by trade union representatives. In addition, many workers are convinced that belonging to an organisation such as a trade union would deny them their freewheeling lifestyle in construction. Consequently, union membership has declined, and trade unions in construction have been forced to amalgamate, continuing a trend begun at the beginning of the century.

Conflict is not unique to the construction sector. Indeed, conflict is endemic in all economies, since it is invariably the case that there are not sufficient resources to satisfy everyone's needs. As a result, giving more to one implies giving less to another. Thus, sharing a contract causes conflicts between main contractors, subcontractors and clients. The higher the price, the better for the seller but the worse for the buyer. Similarly, the lower the wage, the worse for the worker but the better for the employer.

The major negotiating body of the construction sector is the National Joint Council of the Building Industry (NJCBI), which brings together both unions and employers. The Civil Engineering Construction Conciliation Board similarly conducts annual negotiations between the Federation of Civil Engineering Contractors and the main construction industry trade unions. In the Building and Allied Trades Joint Industry Council, the Federation of Master Builders negotiates a separate agreement with the unions. Separate agreements are also negotiated with the National Federation of Demolition Contractors, the Oil and Chemical Plant Constructors Association, and the National Engineering Construction Employers' Association. The Joint Industry Board for the Electrical Contracting Industry carries out similar negotiations with the unions. There are separate agreements in Scotland between the unions and the Scottish Building Employers' Federation. Unfortunately, although various rates of pay and conditions of employment are discussed annually, little of the agreements are reflected on site in practice on a day-to-day basis.

Nor has the construction industry been any more successful in setting up bodies to resolve the conflicts that arise between firms. Firms may go to arbitration, and guarantees known as liquidated warranties may be given, but it remains the case that the construction industry is confrontational and litigious. In 1994, the Latham Report argued in favour of greater co-operation between the members of teams, improved relations between contractors and subcontractors, and improved contract terms.

Turning from the participants in the building process to the suppliers of materials and construction components used, there are several examples of suppliers' organisations, including the Building Materials Producers' Federation, the Builders' Merchants Federation and the National Federation of Builders' and Plumbers' Merchants.

As well as suppliers, there are associations dedicated to the materials they supply or handle: for example, the Aluminium Federation Ltd, the British Cement Association, and the Lead Development Association. Component manufacturers are also represented by their associations: for example, the National Association of Lift Manufacturers, and the Steel Window Association. There are also specialist organisations, such as the Suspended Ceilings Association.

Research into materials and other issues affecting the construction sector is carried out by several organisations, including the Building Research Establishment, the Fire Research Station, and the Transport and Road Research Laboratory. Other research is funded or carried out by several bodies, including the Construction Industry Research and Information Association (CIRIA), the Building Services Research and Information Association (BSRIA), and the Timber Research and Development Association. The Department of the Environment (DoE) also funds research through the Construction Sponsorship Directorate. Finally, the Engineering and Physical Science Research Council is responsible for organising research in construction, as part of the Innovative Manufacturing Initiative, (IMI), set up by government in partnership with industry and commerce.

Testing of materials is carried out by various laboratories operated by manufacturers and some large contractors. Academic research into construction management is carried out in many universities and colleges, and robotics applications in the building process and other issues relating to the building process by groups including the Association of Researchers in Construction Organisation and Management.

Finally, government departments involved in construction include the Department of the Environment Construction Technology Division, the Department of the Environment Building Regulations Division, and the Department of the Environment Construction Policy Directorate. The Department for Education and Employment and the Board of Trade, are also involved in issues relating to construction. The Advisory, Conciliation and Arbitration Service is also available to be called in to intervene in disputes in construction between management and labour. Other public-sector bodies relating to the construction sector include the Construction Industry Training Board, and the City and Guilds of London Institute, which are, of course, concerned with skill training and vocational qualifications.

3 Construction and the Economy

Introduction

To build up a picture of any industrial sector, different aspects of the industry need to be looked at separately. These aspects include the types of goods or products produced, namely the output of the sector, the types of firm producing that output, the different types of labour used, the technology and materials used in the industry, and the types of client who form the demand for the industry's output. This chapter deals mainly with the output of the construction industry for both the public sector or the government, and firms and households in the private sector. The next chapter will deal with firms, their labour and technology.

The output of the construction industry

The construction industry can be divided between building and civil engineering, and firms in the industry erect, repair and demolish all types of building and civil engineering structures. Table 3.1 summarises the output of the construction industry. The most striking feature of the table is the fact that repair and maintenance constitutes almost half of all output of construction firms. Just under 40 per cent of output was new buildings, of which new housing totalled £6628m, while the value of construction contracts for new industrial and commercial buildings amounted to £11 384m. New infrastructure projects, such as roads and bridges, water supply and sewage treatment plants, represented almost one pound in every eight spent on construction. It is the built infrastructure that enables other activities to take place, from industrial production to cultural pursuits. An effective road and rail network, for instance, enables goods to be transported swiftly from factory to customer. Providing the infrastructure of the economy as a whole may or may not be profitable in itself, but it is essential for the smooth running and well-being of society at large.

Since the early 1970s, housing, mainly public-sector housing, has declined in significance to the industry, while repair and maintenance has grown from just over 28 per cent in 1970 to almost half the value of the industry's output

Table 3.1   The output of the construction sector, 1993

Type of work	£m	Percentage[1]
New housing	6 628	14.31
New infrastructure	5 544	11.97
Other new work	11 384	24.58
Repair and maintenance	22 767	49.15
Total	46 323	100

Note   1.   Figures do not sum to 100 because of rounding.
Source: Housing and Construction Statistics, 1994, table 1.6.

by the early 1993. The amount of repair and maintenance work is dependent on the total quantity of buildings, and their age and condition. The greater the quantity and the older the stock, the more repair and maintenance will be required. Also, because of the durability of buildings, the output of new-build projects by the construction sector in any one year will only represent a very small addition to the total building stock, while a continually increasing and ageing existing stock will require additional amounts of repair and refurbishment.

Economic data of the construction sector

The output of the construction sector is divided into housing, infrastructure, industrial and commercial buildings, and repair and maintenance. In order to measure that output, the money value of contractors earnings is used. The figures they report are those current at the time and are called *current prices*. Unfortunately, inflation tends to exaggerate growth and expansion simply because current prices tend to rise from year to year. In order to take out the distorting effect of inflation, economists use *constant prices*. Constant prices are the prices of one year applied to the output of other years. Where money values are used below as a measure of change, the figures are based on 1990 prices similar items.

The statistics below also distinguish between the public and private sectors. The public sector includes local authorities, central government departments and public corporations such as British Rail, owned by the state. The private sector is comprised of all firms owned by shareholders or individuals and households.

Looking at housing output figures in particular, the decline in new house building can be seen following the drive to replace post war slums in the 1950s and 1960s. The drop in public-sector housing since 1968, the peak year for new housing, continued until the early 1990s, but the downward trend appears to have been reversed since 1991, which showed the lowest output over the previous forty years. During the 1980s, annual housing output declined, having risen to a peak for the decade in 1988 (see Figure 3.1).

There was no similar drop in the rate of household formation, which meant that a housing shortage was inevitable. If construction is a response to demand, it can be seen that it failed to meet demand in this instance. The reasons for that failure include the drop in the number of new or extra local authority housing as part of government policy to withdraw from the housing market. At the same time, private house builders failed to increase their building output to keep up with housing needs, because local authorities were in the process of selling off their housing stock, which added to uncertainty in the private-sector housing market. Uncertainty deters investment. Building regulations, interest rates and employment uncertainties all contributed to the shortfall in housing starts by the end of the 1980s.

In fact, the stock of dwellings continued to rise throughout the 1980s, in spite of the drop in annual house building. The reason for this apparent anomaly is that the stock of existing dwellings is static and only the annual output can be used to adjust supply to where it is required. In other words, house builders did not respond sufficiently to changing household patterns and population movements from one part of the country to another. There was no overall shortage of housing. The shortage of housing occurred in

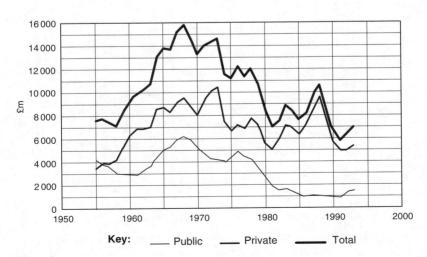

Key: —— Public   —— Private   —— Total

Source: *Housing and Construction Statistics,* 1994, table 1.6.

Figure 3.1   Housing output 1955–93 at constant 1990 prices

specific types or sizes of dwellings and in different areas throughout the country.

During the 1980s output of construction work for the public sector grew gradually from £3.8bn in 1980 to £5.2bn by 1993, while work on the infrastructure rose from £2.6bn to just over £7bn. Private-sector work on industrial and commercial buildings expanded rapidly in the 1980s, largely as a result of the growth of commercial buildings such as offices and shopping centres, only to peak in 1990. Between 1990 and 1993, work on commercial buildings declined by 42 per cent, from £11.3bn to £6.6bn. Although demand for work on particular types of building may be volatile, the figures in Figure 3.2 show that the overall drop in new-build work, other than housing, was just under 11 per cent down from £24bn to £21.5bn, which was similar to the output reached in 1989, the second highest year on record. The reason for the apparent discrepancy between perception and the statistics is because the overall output of the construction industry, which determines the level of employment and profitability peaked in 1990.

We now look at the significance of repair and maintenance on the overall level of demand for construction. Figure 3.3 shows that while housing output in fact increased slightly between 1990 and 1993, and the decline in new-build other than housing fell by 11 per cent, the total demand, including repair and maintenance fell by almost 19 per cent, influenced by the drop in demand for

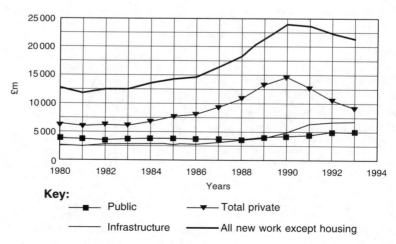

Source: *Housing and Construction Statistics,* 1994, table 1.6.

Figure 3.2   Other new work excluding housing, at 1990 prices

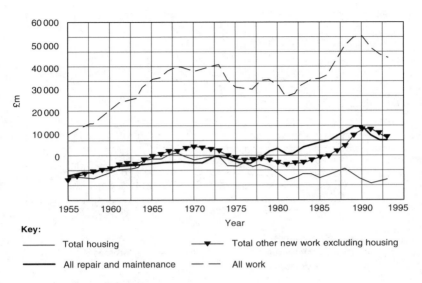

Source: *Housing and Construction Statistics,* 1994, table 1.6.

Figure 3.3   Time series of the composition of total output, 1955–93

repair and maintenance. Total output fell from £55.3bn in 1990 to £48.6bn in 1993, a fall of just over 12 per cent.

Of all the different types of work undertaken by the construction sector, repair and maintenance shows the steadiest, though not the most rapid, growth in the period. From 1978 to 1988, the annual value of repair and maintenance work grew from £10.8bn to £14.7bn, a rise of 36 per cent over the decade. Moreover, repair and maintenance also rose as a percentage of all work undertaken by firms, from 36.6 per cent to 44.1 per cent. Repair and maintenance work depends on the size of the existing stock of buildings, and their age and condition.

Regional variations

The timing of variations in demand varies from region to region, as shown in Figure 3.4, which gives examples of regional variation in the value of new orders obtained by contractors. The total value of new orders has been divided by four in order to indicate the overall trend. It is clear that one region alone, Greater London, accounts for almost a quarter of all new orders, and it is not surprising to see that Greater London closely follows the national trend,

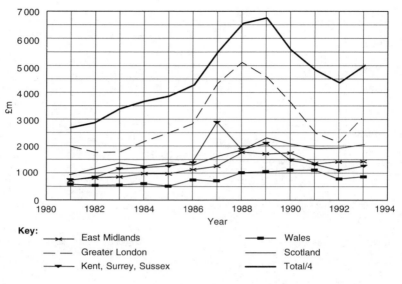

Source: *Housing and Construction Statistics,* 1994, table 1.3.

Figure 3.4   New orders obtained by contractors, by region, 1981–93

because of its dominating statistical influence. As far as the other regions are concerned, there is little pattern in their movements. Although they clearly follow national trends, the timing and size of variations are not possible to predict on the basis of other regions. The size of individual projects plays a more important role in determining the volume of demand in any particular region. For example, the impact of the Channel Tunnel can clearly be seen as influencing the volume of demand in the Kent, Surrey and Sussex region in the period between 1987 and 1990.

The time series of output shown in Figure 3.3 illustrated that variations in different types of work appear to move in the same direction simultaneously. The total output figures of the industry therefore reflect the volatility of demand for different types of building work. Diversifying (that is, spreading into different types of building work) will not necessarily smooth the work-loads of contractors, if all work is increasing or decreasing at the same time. However, diversifying by moving into different regions may help firms cash flows even if it does not help them to deploy labour. They may then be able to take advantage of opportunities in one region while demand for work in another region is depressed. For this reason, firms often merge with or take over builders in different regions as a way of entering particular markets. Construction firms will be discussed in more detail in the next chapter.

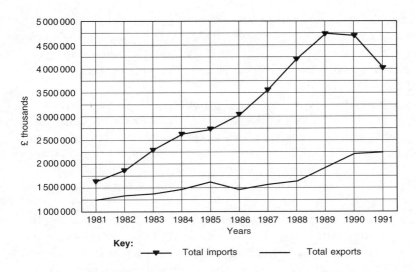

Source: Housing and Construction Statistics, 1994, tables 4.8 and 4.9.

Figure 3.5   Trade in selected materials and components for construction use, at current prices

Exports

The construction industry, like any other industry, trades with other countries. Visible trade involves goods such as glazed and unglazed tiles, materials such as timber, and equipment such as central heating boilers and radiators. Although the value of exports of some categories of goods exceeds the value of imports, taking the value of the range of materials and components used in construction, there is an overall trade deficit; that is, imports exceed exports (see Figure 3.5).

Because construction firms provide a service rather than a product, when they carry out work abroad it was until recently classed as an invisible export. Similarly, when foreign firms undertake UK projects, their contracts are considered to be invisible imports. Figure 3.6 shows work done abroad by British construction firms, including their foreign branches and subsidiaries. The graph illustrates the variation of demand for UK contractors abroad, the decline of the importance of the Middle East as a market and its replacement by the Americas, concentrated mainly in the USA and Canada. In fact, by 1991, North America accounted for all but £57m of the £1015m foreign earnings from the Americas.

Because the graph in Figure 3.6 is based on current prices, the decline in the real value of foreign earnings between 1984 and 1988 was, in fact, greater than shown, and the recovery from 1988 to 1991 not as significant. Current prices rise from year to year in line with inflation, and a rising trend would occur even if no real growth occurred.

In fact, the relative importance of foreign earnings can be seen in Figure 3.7, which shows contractors foreign earnings as proportion of UK construction firms' total output. Since the early 1980s, overseas earnings have declined in importance relative to work carried out by UK building firms in the UK market. Contractors have concentrated their efforts on the home market. After 1989 the recovery of overseas earnings as a percentage of output reflected the decline of the UK market rather than any increase in overseas work.

The construction industry's efforts abroad include designs for foreign clients, the construction of specialised buildings, and turnkey projects where the design and construction is carried out by, say, British or American firms either hiring local labour or sending their own personnel abroad. Although design work for foreign clients may constitute invisible exports, no reliable measure of the value of these exists. Nevertheless, the importance of foreign markets for many UK architects should not be underestimated. The need to find work, the prestige of international competition, and the penetration of the domestic market by foreign architects make it necessary to expand

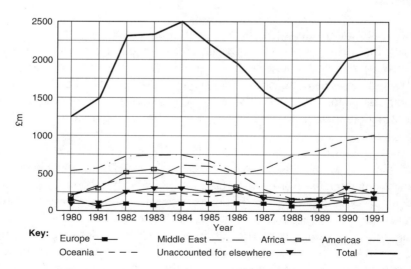

Source: *Housing and Construction Statistics,* 1994, table 1.11.

Figure 3.6   British construction work overseas, at current prices, 1980–91

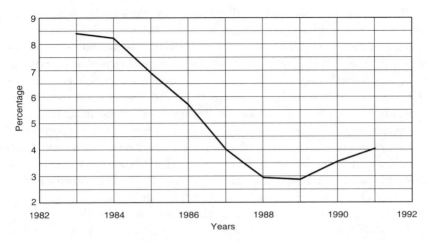

Source: *Housing and Construction Statistics,* 1994, table 1.6; *Housing and Construction Statistics,* 1992, table 1.11.

Figure 3.7   *UK contractors' foreign earnings as a percentage of total output*

abroad. Moreover, in recent years there have been moves by several firms of quantity surveyors to offer their services in various countries in Europe and Asia. Indeed several firms have opened branches abroad.

The economic role of construction

Construction plays an important role in the economy. After all, it produces and maintains the built environment. As we have seen, the built environment consists of infrastructure, commercial and industrial buildings, and housing. These buildings are needed for production to take place and can be seen as investment in the assets of a country. The size and capacity of the construction industry is therefore important in meeting the needs of the rest of society. In order to understand its economic significance compared to other industries, we now look at the size of the construction industry compared to the rest of the economy.

The best measure of the size of an economy is the national income. This shows the total value of all goods and services, including construction, produced in a given year. There are three methods of measuring the flow of goods and services produced: the *output method* calculates the total value of all goods and services produced: the *income method* calculates the total of all incomes earned in return for producing the goods and services; and the *expenditure method* measures total spending (and savings) by all firms, government departments and households.

As all income is earned in return for providing goods and services, the output measure must equal the income measure, and as income can either be spent or saved, the income measure must equal the expenditure measure. Whichever measure is used, the total is the same. The different methods of calculating national income are used to answer different questions. For example, governments need to know about changes in the relative proportions of earned and unearned income before imposing tax changes. Firms may wish to know if consumer expenditure is rising or falling before investing in new plant and equipment. Politicians, journalists and writers need to use information about changes in the economy from one year to the next in order to devise or comment on policies dealing with economic problems.

The sum of the value of all goods and services produced in a country in a given year is called the gross domestic product (GDP). The money value of all goods and services can be measured using their selling prices. These prices are called market prices and include taxes. If the GDP is calculated on the basis of values net of tax, then the prices used are known as *factor costs*. A *factor* in this context means an input into the production process, including labour, materials and equipment. The value of a product at factor cost is therefore the value actually received by the worker or the owner after tax has been deducted.

Whichever measure is used, market prices or factor cost, the GDP represents the actual output of a country in a given year. In 1993, the GDP was £630 023m at market prices and £546 120m at factor cost. However, this is only the part of national income earned at home. There is also a need to take foreign earnings into account in order to measure the purchasing power of UK citizens, but before adding these foreign earnings to the GDP, the profits of foreign owned firms must be deducted. The GDP plus net property earnings from abroad is equal to the gross national product (GNP). In 1993, the GNP was £633 085m at market prices and £549 182m at factor cost. A further adjustment is needed to take into account both money transferred to the UK by individuals abroad and money transferred abroad by individuals in this country. Once these adjustments have been made, a figure for gross national disposable income (GNDI) is found. In 1993, the GNDI was £627 979m.

Finally, the figures for GDP, GNP and GNDI all refer to prices of output at their brand new values. It is, however, more accurate if depreciation is taken into account. Goods produced at the beginning of a year are older than those produced towards the end. Since GNP represents the aggregate of all goods and services produced by the end of a period, an allowance for depreciation, which economists call *capital consumption*, means that the adjusted figure is a more accurate estimate of the total value of output at the year end. GNP less capital consumption is net national product (NNP). In 1993, the NNP at factor cost was £484 159. This is the figure usually referred to as *national income*. The relationships between GDP, GNP, GNDI and NNP are summarised in Table 3.2.

Before interpreting national income statistics, let us be aware of some of the limitations inherent in the data. First, work carried out by individuals in their own homes is not included in national income accounts, since no money transactions are recorded for housekeeping and do-it-yourself repairs. Nevertheless, these activities undoubtedly contribute greatly to the quality of life of the inhabitants of any society. Moreover, a great deal of economic activity takes place informally. It is not declared to the tax authorities, and therefore largely goes unmeasured. However, spending in shops is greater than would be suggested by the figures for reported work, though, of course, some of the spending from legitimate earnings is also spent in the informal sector of the economy. Estimating the size of this sector of the economy is extremely difficult. In the construction industry, under-reporting of work is one of the consequences of the casual nature of employment. Casual labourers can be hired and paid without proper control, enabling many to evade paying income tax.

Prices are used to measure national income, but prices rise and fall, making it difficult to compare one year with another. As national income figures imply a measure of the value of national output, when products such as computers

Table 3.2   *National and domestic product, 1993*

Income aggregates and adjustments	£m
Gross domestic product at market prices	**630 023**
Net property income from abroad	+ 3 062
Gross national product at market prices	= **633 085**
Net transfer income from abroad	− 5 106
Gross national disposable income	= **627 979**
Gross domestic product at market prices	**630 023**
Adjustment to factor cost	− 83 903
Gross domestic product at factor cost	= **546 120**
Net property income from abroad	+ 3 062
Gross national product at factor cost	= **549 182**
Capital consumption	− 65 023
Net national product (national income)	= **484 159**

Source: *National Income Accounts*, HMSO, table 1.1.

come down in price, their real contribution to the economy is understated. Inflation also distorts the GNP statistics, as higher prices do not mean greater output.

GNP does not measure a stock of wealth. It is not a measure of the value of a nation's assets; it only measures the annual income of an economy, which is the value of the flow of goods and services produced in a given year. If income (or the output produced) is greater than expenditure (or what is consumed), there will be an increase in the assets or wealth of the country. In the same way, if individuals earn more than they spend, then their stock of capital will increase. In 1986, the NNP was £288 020m. In the same year, the net value of the total stock of wealth in the UK national balance sheet stood at £1786bn. (In 1086, the year of the Domesday Book, it stood at £73 000.) According to the national income accounts, the stock of wealth is just over six times the national income.

The contribution of the construction industry to national income is complex, since it would appear that the contribution to the GDP takes place only in years of construction, but the benefits derived from construction work stretch over the working life of buildings and structures. Thus buildings not only contribute to the GDP during their construction but also add something to a country's stock of wealth as the stock of buildings in a country accumulates for the benefit of succeeding generations. In fact, in any one year, the output of new buildings of the construction industry only represents a small proportion of buildings offered for sale. Even a small increase in total demand for buildings of a particular type may result in a large rise in the construction of new buildings, if there are not enough existing buildings to meet clients' needs.

The GDP is a measure of an economy's annual income, and economic growth refers to the rate of increase in that income from one year to the next. In any one year, manufacturing industry and construction produce consumer goods and capital goods. While consumer goods are those goods that are bought by households for their own use, capital goods or investment goods are those goods, such as machines and buildings, which add to the productive capacity of the economy. An economy must spend its income on capital goods as well as consumer goods. If the whole national income is consumed on current expenditure, there would be nothing left to invest in the capital goods needed to produce goods in the future. Construction therefore produces both consumer and capital goods, and by providing the latter, it enables the economy to grow.

From Figure 3.8 it can be seen that while GDP grew in real terms from £356 743m per annum in 1971 to £548 559m per annum in 1993, construction only grew from an annual output of £44 190m to £48 554m during the same period. Therefore, construction declined as a proportion of GDP from over 12 per cent in 1971 to just under 9 per cent in 1993. This is shown in Figure 3.9.

For mathematical and statistical reasons beyond the scope of this book, the data shown in Figures 3.8 and 3.9 is not consistent with other data on the construction industry that the government has published. Value-added statistics in the national income accounts, giving the contribution each industry makes to GDP, show that during the 1980s the construction industry only maintained its share of GDP at between 6.0 and 7.4 per cent, and in 1993 was 6.2 per cent.

Nevertheless, from Figure 3.8, the output of the construction industry was sufficient for the rest of the economy to grow between 1971 and 1993. Each year, the construction industry maintained and replaced the stock of the built environment. Indeed, there were sufficient additions to the built stock to accommodate increases in the rest of the economy, although there was no long-run expansion of the construction industry itself. The durability of buildings means that the output of the construction industry is cumulative, provided that the rate of demolition is lower than the rate of building work. As a result, the stock of built environment increases over time. This process is known as the rate of fixed capital formation.

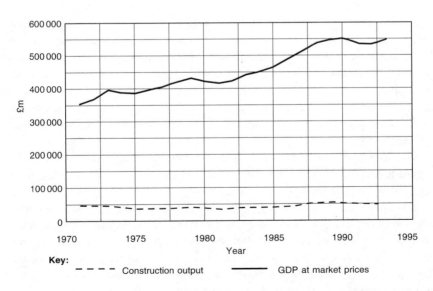

Source: *Housing and Construction Statistics*, 1994, table 1.6; *National Income Accounts*, 1994, table 1.1; *National Income Accounts*, 1993, table 1.3.

Figure 3.8　Construction output and GDP at constant 1990 prices, 1971–1993

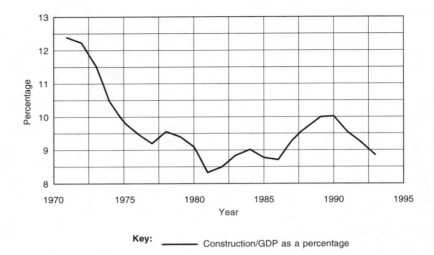

Source: Housing and Construction Statistics, 1994, table 1.6; National Income Accounts, 1994, table 1.1; National Income Accounts, 1993, table 1.3.

Figure 3.9   Construction output as a percentage of GDP 1971–93

An economy's productive capacity is influenced by its stock of fixed capital, including buildings, plant and machinery, and it is therefore important to know whether that stock is increasing or decreasing. In the national income accounts, therefore, tables of gross domestic fixed capital formation (GDFCF) analyse additions to fixed capital in terms of purchases by the private sector, public corporations and general government. GDFCF includes dwellings as well as industrial and commercial buildings. Almost half of the investment in GDFCF is in buildings. In 1991, for example, just over half of GDFCF consisted of dwellings and other new buildings and works, compared to just over a third for plant and machinery. The construction sector therefore plays a vital role in a country's rate of fixed capital formation and economic growth.

Of course, the stock of fixed capital will also depend on the rate of capital consumption. In terms of the built environment, capital consumption means that the building stock is in constant need of repair and maintenance, otherwise buildings become obsolete, dilapidated and unfit for habitation or use. The stock of buildings only increases if the rate of additional new building and repair is greater than the rate of demolition and obsolescence.

Dilapidation and building obsolescence are linked. Buildings become obsolete when they are no longer economic to maintain. This can occur when the rental income is insufficient to pay for the repairs and maintenance. Rentals can decline for a variety of reasons. For example, when an out-of-town shopping mall opens, shops in a town centre lose business and may vacate perfectly sound buildings. These buildings become obsolete, may then fall into disrepair, and eventually become dilapidated.

Economic sectors

The economy can be divided into different economic sectors. Each way of dividing the economy sheds a different light on the role of the construction sector. For example, it is possible to make a threefold division of the economy into primary, secondary and tertiary sectors. The primary sector is concerned with the extraction of raw materials and the harvesting of food from the land and the sea. The primary sector includes quarrying, mining and agriculture, and links society's economic activities to the exploitation or use of the world's natural resources.

The secondary sector includes the manufacturing industries, which produce finished goods from raw materials, including the production of cars, furniture, plant and machinery. The tertiary sector is concerned with the distribution and application of goods. This third sector includes the service industries, such as transport, wholesaling and retailing. Other services include tourism, hairdressing, health, education and governmental administration, legal, banking, insurance and other financial services.

Clearly, construction straddles all three sectors. Several major construction firms possess their own quarries in the primary sector which they use to supply their sites. At the same time, firms in the secondary sector manufacture the prefabricated components assembled on building sites, where more raw materials, cement, aggregates and so on are transformed into finished buildings. Several contractors offer pure management services on a professional fee basis. They do not undertake any of the actual building work themselves, subcontracting all the work packages to specialist builders. The service, management fee contractors provide is by definition in the tertiary sector, as are the services offered by the professional practices of structural engineers, architects and surveyors.

The economy can be divided into 'leading' and 'following' sectors. A sector is a lead sector when an increase of work in that sector causes extra activity in other industries, which are called the 'following sectors'. Construction was used as a leading sector in the 1950s and 1960s in order to stimulate employment and demand in the economy as a whole. Expansion of house building leads in turn to a stimulation of supply industries, such as steel, bricks and building components during construction, and on completion, after a time lag, demand for home furnishings including carpets, furniture and white goods also increase.

The construction industry can also be used to stimulate economic activity in particular parts of the country as part of regional policy. Construction in Belfast became a vehicle for government policy to stimulate employment and reduce tension in the Province. Similar policies have been adopted in Glasgow, Newcastle, and Docklands in East London. Inner city regeneration often concerns policies of refurbishment and construction in run-down parts of cities in order to attract new firms or families into the area. The construction of the Channel Tunnel project can be seen as stimulating the economy in the South East of England and consequently attracting employment into the area.

Sometimes new industries or technologies develop which stimulate the growth of towns or cities, and in those cases, construction is a following sector. The growth of the oil industry in Aberdeen in the 1970s led to an expansion of construction in the city. The building of railway lines led to an increase in construction. For example, the expansion of the London Underground and rail network encouraged the growth of many suburbs. The advent of the motor car also led to increased construction and civil engineering work.

The demand for industrial building depends on growth industries. However, there are periods when industrial development demand can drop significantly. Between 1989 and 1990, demand for industrial building fell by approximately 20 per cent. The role of construction in relation to other sectors of the economy therefore depends on the circumstances at the time. The construction industry can be a lead or a lag sector. Construction expansion is usually necessary, but not sufficient for growth in the economy.

Value added

Every industry uses inputs, such as materials and components, from other industries in order to produce an output. Each firm purchases its inputs at a lower price than it sells its output. The difference is the value added by the firm and value added is therefore the source of income for the people who work in the firm and the derivation of its profits. The total of all value added by all firms is therefore equal to the national income.

Table 3.3 is a summary of the value added by each industry in 1992. According to this table, the share of the construction industry of GDP, as noted earlier, was 6.2 per cent. This was over a quarter of the contribution of the whole of manufacturing industry, which itself was 22.3 per cent of GDP. Manufacturing as a proportion of GDP has been declining for several decades and in 1992 was overtaken by the financial services sector, which earned 23.7 per cent of the national income.

Table 3.3   Value added analysed by industry, 1992

Industrial sector	Percentage
Agriculture	1.8
Coal mining	0.3
Oil and natural gas	1.5
Other mining and quarrying	0.1
Manufacturing	22.3
Electricity, gas and water supply	2.7
Construction	**6.2**
Wholesaling, retailing, hotels and catering	14.1
Transport, storage, etc.	8.1
Financial services, real estate, etc.	23.7
Public admin., defence, etc.	7.1
Education, health, etc.	10.2
Other services	6.4
Adjustment	−4.5
GDP at factor cost	100.0

Source: *National Income Accounts*, 1993, table 16.4.

Table 3.4   Percentage and value of inputs into the construction industry, 1990

Supplying industry	Percentage of total	£m
Mined and quarry products	1.7	946
Chemicals	1.2	646
Iron and steel	2.6	1445
Other metal products	4.5	2450
Mechanical engineering	5.0	2750
Bricks and concrete products	14.4	7849
Timber and furniture	5.2	2864
Air, road, rail and sea transport	0.8	461
Financial and professional services	8.5	4657
Other manufacturers/service industries	5.7	3103
Electrical engineering	1.0	568
Owning and dealing in real estate	4.5	2474
Renting of movables	2.6	1412
Total from outside construction	**57.9**	**31625**
Total from within construction	**42.1**	**22991**
Total purchases	**100.0**	**54616**

Source: *Economic Trends*, no. 480, October 1993; *Input–Output Balance for the United Kingdom*, 1990 (CSO, HMSO).

The relationship between construction and its supply industries in 1990 is shown in Table 3.4. As much as 57.9 per cent of the inputs into construction projects came from outside the industry. Brick and concrete products were the largest category of purchase. In 1990, purchases amounted to £54 616m, including £7849m supplied by the bricks and concrete products industry, £3845m supplied by the metal products sector, and £2864m supplied by the timber and furniture trade industries. In return for the services it provided in 1990, the financial services sector charged the construction industry £4657m.

From Table 3.4 it can be seen that when the construction industry expands to meet demand, it stimulates demand in other industries, because such a high proportion of the output of the construction firms is in fact produced by other industries. Another reason the construction industry may be seen as stimulating the economy, it is often argued, is because it is a labour-intensive industry, and increases in demand for construction create employment, which in turn raises the level of incomes in a region. As people are brought in to paid employment, they in turn have extra money to spend and it is this extra spending which stimulates demand in the rest of society through a process known as the *multiplier effect*.

The multiplier effect

The output of the construction industry can be seen as a series of investments, since decisions to invest in factories and roads are taken with a view to receiving benefits for many years in the future. Current expectations about the future therefore play an important role in the investment decisions taken by firms and governments.

As a result of decisions to invest, labour will be required, creating employment. This employment is the result of decisions made by companies, based on their own optimism about future sales, or perhaps the economy in coming years. Because it is not the result of consumers' expenditure on goods and services, investment decisions inject money into the economy. The money used to pay for the investment then continues to circulate round the economy, passing from one individual to the next as transactions take place. Thus, those who gain employment will in turn spend a proportion of their earnings, and save the remainder. The amount spent on goods and services will create further rounds of new employment, with further increases in income for others in the community. Each time money is passed from one person to the next, a proportion of it will be saved, and the amount passed on in each round will diminish. This process will continue like a ripple effect until the payment for the original investment has been exhausted. The increase in incomes of all

Table 3.5   The multiplier effect

	Earn (£m)	Save (£m)	Spend (£m)
Original investment = £1m			
1st round recipients	1.0	0.4	0.6
2nd round recipients	0.6	0.2	0.4
3rd round recipients	0.4	0.1	0.3
4th round recipients	0.3	0.1	0.2
5th round recipients	0.2	0.1	0.1
6th round recipients	0.1	0.1	0.0
Total	**2.6**	**1.0**	**1.6**

the recipients will in the end be greater than the amount of the investment. This increase in incomes forms a measure of the effect of an investment in the construction industry on the rest of the economy, and depends on the value of the multiplier effect. The multiplier is shown schematically in Table 3.5.

Thus, if the increase in all incomes = £2.6m, and the original investment = £1m, then, original investment × the multiplier = increase in all incomes. Substituting values in Table 3.5:

$£1 \times$ the multiplier $= £2.6$m

Therefore, the multiplier $= 2.6$

To put the same thing in mathematical notation: Δ means 'change'. Thus ΔY means change in total income and ΔI is the difference in total investment caused by investing in any one project.

if $\Delta Y = £2.6$m, where $\Delta Y =$ the increase in all incomes   (3.1)

and $\Delta I = £1$m, where $\Delta I =$ the original investment   (3.2)

If $\Delta I \times k = \Delta Y$ (3.3)

then $k = \Delta Y / \Delta I$ (3.4)

and substituting Equations (3.1) and (3.2) in (3.4)

$k = 2.6$ (3.5)

Thus investment expenditure will have a multiplier effect on the economy; namely, the effect of investment is to increase national income by an amount greater than the original investment.

The average propensity to save (APS), is the ratio of total savings to total income, measuring the proportion of income saved. The marginal propensity to save (MPS) is the ratio of the change in savings to a change in income. For example, the MPS measures the proportion of an increase in income that is saved. In the above example, since there would have been a certain level of income before the investment, the effect of the investment is to raise incomes in an area. It is therefore the MPS that reveals the change in income and savings. It will be noted that the increase in incomes, derived from the original investment above, was £2.6m, and the increase in savings was £1m. Therefore;

$$\text{MPS} = \frac{£1\text{m}}{£2.6\text{m}} = \frac{1}{2.6} \tag{3.6}$$

However, the multiplier in the above example was shown to be 2.6. In fact, the multiplier is the reciprocal of the MPS, namely 1/MPS. Thus,

$$\text{if MPS} = \frac{1}{2.6} \tag{3.7}$$

then, the multiplier

$$= \frac{1}{\frac{1}{2.6}} = 2.6 \tag{3.8}$$

This relationship reveals that the impact of an investment in an area will ultimately depend on the willingness and the ability of recipients of investment expenditure to save or consume their extra earnings. The higher the proportion of extra income saved, the lower the multiplier effect.

The impact of the multiplier effect will be reduced not only by savings but also by taxation and spending on imported goods, since these factors reduce the amount of money left for transactions that create employment or raise income in the domestic economy. Moreover, if one takes the locality of the construction site alone, the multiplier effect will be minimal, as the vast bulk of any earnings will be spent further afield. Many beneficiaries of increased earnings in one part of the country may well live in another region or country, in which the extra consumer goods are manufactured. The multiplier is thus often reduced to a figure approaching one, which means that, apart from creating employment for those firms and employees engaged on a project, little further economic impact, such as increased employment, will be felt as a direct result of the construction itself. However, if factories are built or visitors are attracted into an area as a result of a new addition to the built environment, such as a theme park, then employment will be generated, but such employment is not the result of the multiplier effect: it is a consequence of the new completed structure itself.

The accelerator principle

It is necessary to distinguish between net investment and the investment needed to maintain goods in any given year, equivalent to depreciation. If a house, for example, is not properly maintained, it will depreciate like any other asset. The asset only holds its value if it is maintained, assuming everything else remains the same.

Some investment in equipment or buildings may be necessary because they wear out over time, depreciate and need to be replaced. Total investment less replacement investment is net investment. The theory of the accelerator principle attempts to illustrate the effect of national income in one period on the level of net investment in the next. This can be shown as;

$$I_t = v(Y_t - Y_{t-1}) \tag{3.9}$$

where

$$I_t = \text{current net investment}$$
$$Y_t = \text{current national income}$$
$$Y_{t-1} = \text{national income in the previous period}$$
$$v = \text{the accelerator factor}$$

Thus the accelerator is a constant fraction of the change in national income from one period to the next. Changes in national income influence business decisions taken by industrialists. Apart from changes in national income, stock levels held by firms and expectations about future growth also influence investment decisions.

Final goods are products such as shoes, television sets and cars, which consumers commonly buy in retail outlets for their own personal use. Intermediate products are goods that are part of the production process leading to final goods; these include factory machinery and indeed the factories themselves. The accelerator principle relates changes in the demand for final products to changes in the demand for investment goods. Hence, if the rate of decline in demand for, say, television sets, slows down the manufacturer's own demand for equipment to make the sets may be altered drastically. The firm now needs to make new investment plans if only to order replacement machines. These machine tool orders increase employment in the firms providing the equipment. This increase in demand for tools may come about not as a result of an increase in the demand for television sets but because their rate of decline has been arrested and the firm then wishes to maintain its current output. The accelerator therefore relates the performance of the television manufacturer to the machine tool supplier.

At the other end of the trade cycle, when a manufacturer no longer has to buy extra machines to accommodate a growth in sales of television sets, s/he will require fewer machines. No machines will be required for increased capacity. The only equipment required when sales remain at the same level as the previous year will be replacement machinery for those machines that have reached the end of their useful or economic life. Hence, when the rate of increase in demand for televisions slows down, the demand for the machinery to make them declines. Thus, the slow-down in the rate of increase in a final product has a decelerating effect on the demand for the machinery. Again, this has employment implications for the firms supplying the machinery.

The construction industry also provides intermediate investment goods. These are the factories, shops and offices, that are built, not for their own sake, but for the goods and services, that they will provide in succeeding years. Companies often invest in buildings just as they invest in machinery, when they see opportunities for growth. Investment takes place when the economy is expanding and sales are on a rising trend. Eventually, when markets become saturated, that is when it becomes increasingly difficult to sell extra units of a product. Even when sales are at a high level, it becomes difficult to see further opportunities for expansion, at least in the domestic market. A company at the top of its trading cycle may have no investment projects in hand, at least in the domestic market, especially when it may be able to direct investment funds abroad to a country with growth potential or higher rates of return.

When the economy appears to be booming, industrialists often see little room for expansion. There may be bottlenecks and shortages of certain raw materials or skilled labour. Stock inventories may be excessive and firms may attempt to de-stock by reducing the rate of production. Far from replacing buildings, firms are happy to consolidate or even contract.

This will mean fewer new buildings are required at home, as little growth in the economy is anticipated. Firms may then look abroad to fulfil their expansion plans. This drop in investment at home in spite of record sales will become part of a self-fulfilling prophecy. As investment in buildings and new plant and equipment declines, unemployment rises. The newly unemployed are no longer able to purchase the quantities of goods and services that had previously been within their purchasing power. This reduces demand in general, and companies respond with further cuts in investment in plant, machinery and factory buildings. They also reduce their inventories of stock held in warehouses. All this leads to further rounds of reduced investment, particularly in new buildings. Employers in the construction industry have no alternative but to reduce the number of employees to reduce staff costs.

This will continue as a pattern for a number of years as the economy declines, but there will come a point, when even though sales are weak, repairs which have been delayed must be carried out, if only in order to remain in business. Gradually confidence and optimism return. Improvements in profitability and a halt to the rate of decline even at low points in the trade cycle provide industrialists with opportunities for expansion once again. Firms will find they have little stock in reserve. They must manufacture to order. As extra orders are placed, there is a need once again to expand. Extra labour may be taken on to satisfy demand. Although manufacturing industry itself may have had relatively few orders, there may be a rapid increase in the amount of office and factory building taking place, in anticipation of future increases in demand. This increase in work for the labour-intensive construction industry will, in turn, generate more employment and increase demand and optimism in the rest of the economy.

The accelerator principle accounts for the downturn in the economy at the top of the cycle and the upswing after a period of recession. However, the accelerator implies large fluctuations in the demand for industrial building, but the fluctuations are not as great as might be expected because it is often possible to take up some slack in a firm and make use of surplus capacity, perhaps by shift work or overtime. Alternatively, better use of the space available will often enable firms to expand output without resorting to finding or building larger premises. Technological change and innovation in factories and offices often saves space. A rise in demand may induce firms to invest in new machinery to increase output, but achieve it in less space. Thus, although many of the conditions for an upturn in demand for buildings may be present, it does not necessarily take place in the construction industry even when other sectors of the economy are thriving.

Construction as a tool of government economic policy

Government policies related to the construction industry are concerned primarily with meeting society's needs on the one hand and the level of employment on the other. To begin to deal with these issues, the government needs an understanding of the capacity of the construction sector. Capacity is the ability of the construction sector to respond to the demands made upon it. The capacity of the construction sector will itself impinge on government policies. For example, the greater the proportion of construction output which uses imported building components, the greater the effect of building on international trade balances and exchange rate management. If the construction sector is operating at or near full capacity, any inability of contractors to meet demand will raise tender prices and in turn result in inflationary pressures.

Partly because of the multiplier effect, the construction industry has often been considered by governments as a suitable vehicle of economic policy. Such policies are known as 'fine tuning', because of the stimulus the government can give to the whole economy or even to specific regions by, for example, generating work in the building industry. Not only will people be employed in the construction industry, but those construction workers will also create work for others as they go out to purchase goods and services for themselves. Furthermore, whenever factories, offices, shops or houses are built, industries which provide goods used in conjunction with buildings, (for example, furniture, carpeting and machinery) will also be stimulated, as demand for these goods complements the demand for the buildings themselves. The extra jobs created in other industries should be borne in mind even though the cost of creating one job in construction may appear expensive in itself. Moreover, from the government's point of view, people in employment pay income tax, while unemployed people receive social security benefits. Such policies could be seen in 1963, for example, when unemployment reached relatively high levels that were politically unpopular. This spurred the government on to introduce measures to stimulate the economy, and especially the construction industry. By the second quarter of 1964 the volume of new building work was 48 per cent higher than it had been on average in 1958. In a period of two years, factory approvals had risen from 8.1 million square feet to 13.6 million square feet. The number of houses being built in the UK in that quarter was 432 000. By July 1964, according to Lewis (1965), the stock of bricks, which amounted to only 77 million, was equivalent to the output of about three days, while stocks of cement came to less than a week's supply.

By 1971, as much as 56 per cent of building work was carried out on behalf of the government, local authorities and public corporations. Moreover, much of the work carried out in the private sector would also have been in response to government orders for supplies given to firms. Clearly, the government was in a position to control the level of work in the industry and in that year it introduced an emergency public works programme, lasting two years, to combat unemployment and stimulate the economy.

However, there are several difficulties associated with policies designed to stimulate activity in the economy to compensate for a lack of demand in the private sector. First, the execution of such policies may take a long time, often years. Land must be acquired, buildings designed and there may even be a public enquiry for large and controversial projects. By the time construction actually begins, the original reason for the policy — namely, a high rate of unemployment — may no longer apply. Another difficulty in justifying such policies is that if a project is worth constructing, then delaying it indefinitely to time its construction with a period of high unemployment deprives society of the benefits of the building while adding to its costs.

There is also the short-term problem of the government crowding out the private sector as both compete for scarce resources. If this occurs, then the government will only have succeeded in replacing work in the private sector with work in the public sector without in fact increasing the total output of the industry. This will tend to fuel inflation as building costs and tender prices are raised in response to high demand.

More recently, the government has reduced the size of the public sector through a programme of denationalisation or privatisation, and by reduced government intervention in the economy and in particular markets. This policy is most noticeable in the housing market, where the sale of council housing to sitting tenants has been accompanied by the rapid decline in public-sector house building.

The Keynesians versus the monetarists

Behind these shifts in policy lie differences in economic theory, to which governments of the day have adhered. For a period stretching from the end of the Second World War until the mid- to late 1970s, the ideas contained within the Keynesian approach held sway. Even before 1979, however, when the Conservatives won the general election, a monetarist view of the economic causes of inflation and unemployment gained ground and influenced government policy and actions. As a result, the government has continued a policy of reducing public-sector spending as a percentage of the national income.

Perhaps the fundamental difference between Keynesians and monetarists is the belief that markets have a tendency to clear. In other words, monetarists maintain that if there is a surplus, prices will drop until the surplus has been sold off, whereas Keynesians point to persistent unemployment as evidence that, in the real world, markets do not operate smoothly without government intervention.

In his *General Theory*, in the 1930s, Keynes argued that the economy, if left to itself, would not generate sufficient demand to maintain full employment. People tend to save some of their income and this causes reductions in the level of demand in the economy. Unless the government intervenes to make up the difference, unemployment will persist. Thus, if it wishes to maintain full employment, the government is charged with the pump-priming duty of placing orders for buildings, roads or defence equipment, even though tax revenues may be insufficient to cover the cost. Of course, Keynes recognised that such policies could well lead to inflationary pressures, which would, however, be alleviated to some extent by increased output that would otherwise not have taken place. Stimulating the economy makes more goods available in the marketplace, which increases competition and tends to keep price rises to a minimum. However, other economic forces are also at work, and increased demand for raw materials will raise raw material prices, which will in turn tend to raise the prices of finished goods.

Monetarists, such as Milton Friedman from the University of Chicago, argue that boosting demand by increasing budget deficits (when governments spend more than they take in taxation) only increases the governments' debt and borrowing needs. This increase in government spending increases the money supply and hence inflationary pressure as more money is available to pay for a given quantity of goods and services. Just as in an auction, prices will be bid upwards if people have money in their pockets, so prices offered in shops will rise. Hence the level of prices depends on the quantity of money in circulation. This relationship between the quantity of money and the level of prices has given rise to the Quantity Theory of Money, a theory with roots in the eighteenth century. It was restated by Irving Fisher, an American economist, at the beginning of the twentieth century.

Fisher's equation of exchange stated that $MV = PT$, where $M =$ the quantity of money, V is the rate at which money circulates round the economy, passing from one person to the next at a certain velocity, P is the average price of all goods and services, and T is the number of transactions taking place in a given period of time. Therefore, the total of money changing hands is equal to the number of transactions multiplied by their average value:

$$\text{If } MV = PT, \tag{3.10}$$

$$\text{then, } M = \frac{TP}{V} \tag{3.11}$$

$$\text{If } T \text{ and } V \text{ are constant, then, } M = aP \tag{3.12}$$

where $a =$ constant

Equation 3.12 shows that if the number of transactions and the velocity of circulation are constant, then an increase in the quantity of money must lead to an increase in the price level.

A revised version of the equation of exchange, the Cambridge Equation, further refined the quantity theory. Transactions such as social security payments do not contribute towards the national income, which is a measure of the country's productive capacity. Instead of PT, which includes all transactions, let Y represent those transactions which consist of the goods and services produced in a given period, multiplied by their average price. Hence the Cambridge version stated $MV = Y$ (though the definition of V, which accounted for all transactions in the original formula, was modified). This formula related M to Y. The stability of national income was therefore related to the stability of the money supply.

This view of money assumes that all money is used for transacting business. However, Keynes pointed out three reasons or motives for holding money. Certainly, the first reason was indeed the transactions motive. Then came the precautionary motive, which reasoned that people held a certain amount in cash in order to cope with unexpected emergencies. Third, Keynes argued that people held money for speculative reasons, buying and selling government bonds in response to interest rate changes.

These reasons for holding money broke the close link between money supply and national income. Keynes argued that the quantity or supply of money and the demand for money, which he called liquidity preference, determined interest rates. Interest rates in turn influenced investment, and investment would raise national income more than the amount invested because of the multiplier effect. However, this would not guarantee full employment, according to Keynes, especially as prices and wages do not necessarily respond to changes in market conditions. In summary, the quantity of money influences interest rates rather than national income, prices or employment.

It may well be the case that under certain circumstances one or other of the two conflicting theories better serves to influence policy. Clearly, there may be times when high rates of inflation require governments to take monetarist measures to control price rises. At other times, the more appropriate strategy may be to increase the number of job opportunities. It is not the purpose of this book to determine which of these policies should be adopted.

Some may claim it as an achievement of monetary policy in the 1980s that the rate of inflation was reduced from over 25 per cent per annum to under 5 per cent, while others may accuse the policy of raising unemployment levels to around 14 per cent, with large regional variations. It is, however, important to note that both of inflation and unemployment have largely followed international trends. While the UK rate of inflation did drop, inflation rates were dropping in the rest of the world too, and the relative position of the UK compared to its commercial and industrial rivals did not alter significantly. It is also true to say that unemployment rates were rising in other countries in Europe and North America in the late 1970s and early 1980s. The British unemployment rate rose more than the rates of Britain's main competitors.

The shift to monetary policies in the late 1970s and most of the 1980s, led governments to attempt to control inflation through the money supply. Monetarists advised the government to control the money supply through raising or lowering interest rates. Hence interest rates were raised to historically high levels in an attempt to control sterling M3, the measure of money supply, which the government adopted as the most appropriate monetary aggregate to determine the success of its monetary policy. Sterling M3 was basically the total of all UK currency held in cash and in current (cheque book) accounts in banks in the UK. The term M3 distinguishes this aggregate from others such as M0, which is purely coin and notes in circulation.

These economic policies and consequences were accompanied by a shift in political values, often referred to as Thatcherism, from state help and intervention to self-help and individual responsibility. More reliance was placed on market forces. The idea of the corporate state controlling the lives of the citizens has been replaced by the idea of the private entrepreneur, whose self-interested pursuit of profit can only succeed if satisfying the economic needs of the community. In fact, in modern economics, the arguments between Keynesian and monetarist views have largely been replaced by more empirically based (and realistic) theories on how markets actually operate, transaction costs and the behaviour and decision-making processes within economic actors or participants, such as households, firms and governments. The discussion has shifted to issues, including the social structures of accumulation, which describe the institutional working of markets, including the legislative constraints and the traditional methods used for trading as well as the global changes affecting production, distribution and consumption. These are topics which lie outside the scope of this book.

Government policies

Government policies concerning the construction industry aim to achieve two main objectives; first, construction is needed to meet both the public and private sectors' requirements for a built environment. Second, the construction sector can be used to create employment in a particular region or in the economy as a whole.

To begin to deal with these issues, the government needs an understanding of the capacity of the construction sector. Capacity is the ability of the construction sector to respond to the demands made upon it. The capacity of the construction sector will itself impinge on government policies. For example, the greater the proportion of construction output that uses imported building components, the greater the effect of building on international trade balances and exchange rate management. If the construction sector is operating at or near full capacity, the inability of contractors to meet demand will raise tender prices and in turn result in inflationary pressures.

Economic planning in the construction sector is concerned with growth of capacity in line with growth in demand. Industrial capacity can be increased by increasing the quantity of machinery used per worker, and by education and training. Increasing the output of construction also depends on innovation and technical change. This last point is one of the main functions of management in the construction sector. Namely, the management of change through the introduction of new technology to improve productivity, pay and profits.

There are many problems involved in measuring capacity, which is a constantly changing figure, because of the continuous introduction of new technology. In any case, the capacity of an industry is its maximum potential output, rather than the output actually produced. It is therefore not possible to measure capacity accurately. For example, many people who are unemployed or working in other industries but with experience in the building sector would not necessarily wish to return to the construction industry at any wage. Moreover, the supply of extra labour from Ireland and other countries ensures that building projects can always be started, at a price. Similarly, when materials and components are in short supply, they too are imported, as they have been especially during the early 1990s.

When considering capacity, it is also important to take demand for construction work into account. Construction demand is cyclical. Periods of growth are followed by periods of decline or recession. Recent major recessions in construction occurred in 1973–4 and 1980–2. From 1983 until the end of the 1980s construction demand grew, especially in the commercial property sector. Since 1990 the construction industry has failed to regain the level of demand it achieved in the boom years of the 1980s. Capacity is determined by the ability of the rest of the economy to consume the output of the building industry. Since the 1990 the construction sector has clearly experienced excess capacity, with many firms going into liquidation and construction workers becoming redundant.

If demand falls and firms are forced out of the industry, productive capacity declines. The resulting unemployment would also lead to the departure of skilled and experienced craftsmen from the industry. In 1978, a report published by the National Economic Development Office, entitled *How Flexible is Construction?*, argued that the extent of spare capacity that firms maintain in a downturn will be influenced by their assessments of their *normal* level of demand: if they perceived that demand had declined permanently, they would not wish to keep all their staff. By the 1990s it was no longer a matter of perceiving a permanent decline before shedding labour, but more a case of insufficient work even in the short run.

This is particularly important during recessions in construction demand. A period of adjustment to low levels of demand occurs as firms go to the wall and skilled labour withdraws from the construction sector, through unemployment or finding alternative work. Eventually, following the period of adjustment, the reduced capacity of the industry matches the level of demand, and the fewer firms and workers remaining in the sector begin to increase their workloads, though the industry as a whole is smaller than before.

It can be argued that firms in the construction industry, such as general contractors, are always underutilised. Their ability to supply construction services invariably exceeds effective demand for their output. After all, as capacity is approached, tender prices and final building costs would tend to rise; construction times and waiting lists would tend to lengthen; delays and inefficient working would result; and quality would tend to suffer. Excess capacity may therefore be seen as being necessary in order to maintain steady prices, and minimise the time lag between an increase in demand and an increase in supply.

The employment strategy of a majority of contractors is to minimise the number of directly employed workers, provided that there is a relatively large number of skilled and semi-skilled labour available to take up casual employment. How large the latent pool of skilled labour available to the construction industry is, will vary depending on the economic conditions in the labour market in the economy as a whole. Moreover, many workers currently working in the industry may not be registered and are hence not recorded in the statistics. Much of the work they carry out is also unrecorded and the extent of the black economy in the construction industry can only be guessed at.

Once a contract has been landed, contractors obtain casual labourers, using subcontractors to save on time and administration costs. It is this separation of contractors from the labour force which is at the heart of many of the difficulties currently facing the construction sector. For example, this separation does not encourage contractors to train their temporary workforces fully, and employment is offered only when work is available. If tenders are not won, contractors do not pay for casual labour, and the unemployed workers then receive state benefits.

According to Ive (1983), excess capacity in construction appears as unemployed construction workers, under-utilisation of hire firms' plant, and a flow of money-capital from construction to other sectors. An indication of the relationship between output and capacity in the construction industry was given in *How Flexible is Construction? published by NEDO in 1978*; in 1976, the construction industry had been operating with 30 per cent excess capacity.

However, it is not possible to reach a conclusion about the capacity of the construction sector without making several assumptions. For example, the capacity of the industry depends on the organisation and ability of the management of the individual firms. One would also need to take into account the state of technology used, the methods of construction, the condition of plant and machinery, levels of stock of materials, and the availability as well as skills of the employed, self-employed and unemployed workforce. A major constraint on the ability of contractors to carry out projects is their financial strength at any one time. Finally, the construction industry responds to demands made upon it. If demand increases, then the sector responds by importing materials, components and labour. However, it is always possible to increase the quantity of the built environment by lowering standards and building specifications. In the following chapter, the organisation and capacity of the construction industry will be considered in more detail.

4 Supply and Demand in Construction

Introduction

The price of any transaction agreed between buyers and sellers reflects broad as well as specific forces at work in the economy. These economic forces must be taken into account for a proper understanding of why prices are as high or as low as they are. This chapter deals with the price mechanism, or the laws of supply and demand.

Economic models of the construction industry, like all models, are simplifications of the real world. They are a combination of mathematics and words used to understand the relationship between economic cause and effect in the construction industry. In order to do this it is necessary to ignore temporarily many complications which arise in the construction industry, such as the different types of buildings, different services offered, regional variations, the different objectives of individuals and firms, and constantly changing circumstances. In order to understand the processes involved, economists argue as follows: assuming that everything else is held constant, if one variable changes, then it is possible to isolate it as the cause, and show the effect on, say, the quantity demanded. This method of simplifying the real world is used to build economic models. These consist of manageable statements designed to understand the real world and to help predict events. Later, some of the restrictive assumptions used in the model will be relaxed to make our understanding more realistic.

First, this chapter will analyse the factors influencing demand, then those influencing supply. Having discussed demand and supply separately, the inter-relationship between these forces will be described. This will be followed by a description of different types of market in which the forces of supply and demand operate. Finally, some complicating factors such as inflation and unemployment will be taken into account.

Building activity is determined by the interaction of property prices, interest rates and building costs. This relationship was appreciated as far back as 1877, when The Economist's *Commercial History and Review* of that year referred to the low rate of interest and high demand for property, which encouraged developers to expand the supply in excess of what the market could sustain. Why did developers build more than they could sell? How do markets operate? We shall return to this issue again in Chapter 9, where construction markets are discussed, but in the meantime, we begin our answer to this question here, with the laws of supply and demand, taking property markets as our example. A market price is arrived at through the interaction of the opposing forces of demand and supply. Several factors, called variables, influence the demand side, while a similar, yet distinct, set of conditions determine supply. To understand the market forces at work, a systematic approach is called for, and each variable will be examined in turn.

Effective demand

If demand is based on desire, the question arises, what turns the wishes of individuals into the goods and services, and the possessions they enjoy and have at their disposal? What makes this demand effective? Effective demand is the quantity of goods or services people are willing and able to buy in a given period of time and at a given price. In the construction industry, effective demand is the quantity of building services and projects clients wish to buy. Effective demand is demand backed by purchasing power. Without access to cash in one form or another, demand would be ineffective.

Clients may come from the private or public sectors, and there is a further useful distinction that can be applied to the clients of the building industry: the developer-dealer client wishes to build in order to sell on to a final user, and the non-dealing developer commissions work for his or her own use, for owner occupation. This type of client may be a company needing a factory, a private household wanting a house, or a public authority wishing to build a facility for the use of the general public, such as a museum or a hospital.

The demand for buildings may be seen as a form of investment as the products of the building industry tend to take one or two years to construct, whereas the benefits which flow as a result may last for many decades. This applies equally to factories which enable future production to take place, roads which enable future goods to be transported efficiently and houses which generate a return in the form of greater or lesser comfort for the occupants in years to come.

Derived demand

The demand for construction services is said to be a derived demand. Products such as cars, computers and cornflakes are final goods. They can be purchased by consumers. The plant and equipment and the materials needed to manufacture final goods are called 'intermediate goods'. The demand for intermediate products depends on the demand for final goods. The demand for factory buildings is derived from the demand for the products made in factories, just as the demand for insurance offices is derived from the demand for insurance. This places the demand for many construction services at least two removes from the final customer. It is necessary to take this relationship between construction and final goods and services into account in order to anticipate the workload of construction firms and professional practices.

The quantity of any particular building type demanded will depend on a combination of the following factors.

The price

The price of a building is perhaps the most important single factor influencing demand for property. Assuming everything else stays the same, the higher the price, the less will be demanded; the lower the price, the more will be demanded. This is known as the first law of supply and demand. The relationship between price and the quantity demanded can be seen in Figure 4.1, which represents the demand for homes of a particular size and quality in a given locality. Thus at $£P_1$, a builder may assess the number of homes that might be demanded at Q_2 units; at $£P_2$, only Q_1 units. If these figures are then plotted on a graph, the line connecting such points is known as a *demand curve* and thus aids the prediction of demand at various other prices, always assuming that the circumstances remain unchanged.

The downward-sloping demand curve shows that when price, and only price, changes there will be a change in the quantity demanded. To understand the use of this diagrammatic technique, it is important to remember that a change in price, and only price, leads to a movement along the curve. The reason for stressing this point will become obvious in a moment. Besides price, many other factors influence the quantity buyers are willing and able to purchase.

Interest rates

Interest rates have a key role to play, especially in property and construction markets, because these markets are heavily financed through borrowing. High rates not only deter people from borrowing in order to purchase property, they also make it non-viable to invest in developments with lower rates of return. Thus high interest rates will depress both property prices and the profits that can be made out of developing land. This in turn will have an adverse effect on the demand for the services of builders and building professionals.

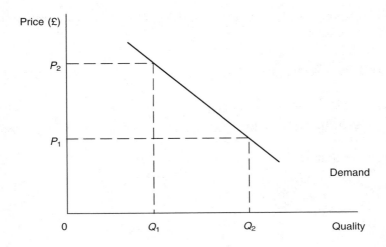

Figure 4.1   Price and the quantity demanded – the demand curve

It is important to distinguish between nominal and real rates of interest. Nominal rates are the rates quoted by banks or lenders. The real rate of interest is approximately equal to the nominal rate of interest minus the rate of inflation. In November 1987, for example, the nominal rate of interest stood at 9 per cent and the rate of inflation was 4.5 per cent. Hence the real rate of interest in the UK was 4.5 per cent. This figure can be compared to rates in other countries such as West Germany (before unification), where inflation was below 1 per cent, and the real rate of interest at the time was as low as 2.6 per cent. At the same time, the real rate of interest in the USA was 4.45 per cent.

Assuming an inflation rate of 3 per cent p.a., prices would rise from £100 000 to £103 000 and £106 190 after two years. If the nominal rate of interest was 8 per cent, then capital would appreciate from £100 000 to £108 000 after one year, and to £116 640 by the end of the second year. The growth in purchasing power, however, increases faster than the real rate given by simply subtracting the rate of inflation from the nominal rate of interest, in this case 8 per cent less 3 per cent. A real rate of interest of 5 per cent would imply an extra £5000 of purchasing power after one year and an extra £10 450 after two years. However, these figures overstate the increase in purchasing power. The true rate of increase in purchasing power must take into account that future additional money units will themselves have diminished in purchasing power because of inflation. When the sums of money are as large as those involved in building projects, the more accurate formula used to find real interest rates is:

$$\text{Real interest rate} = \frac{1 + r}{1 + I} - 1 \tag{4.1}$$

where r = the nominal rate of interest
and I = the rate of inflation.

In the above example the real rate of interest is therefore:

$$\frac{1.08}{1.03} - 1 = 0.04854, \text{ or } 4.854 \text{ per cent.}$$

Thus, after the first year real purchasing power is up by only £4850 and after two years, £9940 instead of £5000 and £10 450, respectively. This difference becomes important in property and construction where projects are frequently in excess of £10m.

It is the real rate of interest that will influence the demand for the construction industry's products. Even when nominal rates of interest are high, they can be met, as developers are in a position to pay the high rates of interest by raising their prices, rents and so on in line with the rate of inflation. Interest rates play a significant role in determining the economic viability of projects, since the rate of return on a project must be greater than the cost of borrowing. Interest rates are also used to discount future costs and benefits in feasibility studies and investment analyses, as will be seen in a later chapter. Assuming that everything else remains the same, an increase in interest rates reduces demand for construction, while a reduction in interest rates will lead to an increase in demand.

The prices of other products

Demand also depends on the price of other products, some of which will be substitutes, some complements and some independent goods. Substitute goods are those that can be used instead others, while complementary goods are those used in conjunction with each other. Independent goods are those that have no direct relationship with each other. An example of substitutes in the construction industry are the services offered by firms bidding for the same project. They will, in effect, be substitutes for each other, and the demand for each firm will be affected by the tender prices offered by competitors. Similarly, wooden window frames are substitutes for metal ones and vice versa, and the price of one type of frame will affect the quantity demanded of the other.

Complementary goods are goods that are bought jointly: for example, doors and door handles. Oil and oil-fired central heating systems are also complementary goods. An increase in demand for one leads to an increase in demand for the other. Similarly, a fall in the price of one leads to a rise in demand for the other. For example, if the price of petrol goes down, more petrol will be demanded. Motorists then increase the use of their cars and wear down their car tyres in the process, and this increases the demand for replacement tyres. Tyres and petrol are therefore complementary goods. Similarly, the more work given to architects, the greater the demand for quantity surveyors. Their services are said to be complementary. In a sense, they are interdependent. The greater the demand for houses in a given area, the greater the demand for utilities, shops and transport facilities to service them.

Of course, not all goods have a substitute or complementary relationship with each other. The price of salt does not directly affect the demand for furniture. These goods may be said to be independent of one another. In the nineteenth century the house was one of the few durable items a family could spend its money on. As the twentieth century has progressed, more and more products and services have begun to emerge as major drains on a family income: cars, holidays abroad and a whole variety of electronic gadgetry have become generally available. Nevertheless, the price of these products, though independent, affects the amount consumers are willing to spend on accommodation, and vice versa. Taken together, these products compete for the consumers' limited resources. Thus demand for a particular product or service is affected by the price of other goods, be they substitute, complementary or independent.

Income

As stated earlier, effective demand is dependent on the ability to pay, which is in turn dependent on income. Unless the government intervenes, low income is a major cause of inferior housing for many individuals. Indeed, this issue of low income lies at the heart of the problem of poor housing and homelessness, even at a time when many construction workers are unemployed, but capable of building dwellings.

Assuming that everything else remains constant, a change in income will alter an individual's effective demand. Economists distinguish between two types of good, normal and inferior. Normal goods, such as silk scarves or claret, or good quality finishes on buildings, are those which consumers tend to purchase in greater quantities, the higher their income. Demand for inferior goods, such as sausagemeat or second-hand cars or poor-quality finishes, will decline as incomes rise above a certain level, when people begin to select a superior quality product.

The quantity demanded can, of course, be plotted against different income levels. Such a graph is known as an Engel curve, after the economist Ernst Engel. An example of a normal good is illustrated in Figure 4.2. As the income of consumers rise, so does their demand for certain goods and services, such as holidays abroad, cars, clothing, and also the size of dwellings.

However, certain inferior products may not follow this pattern. Demand for them will rise to begin with as incomes rise but then, as incomes continue to rise beyond a given level, demand will decline. This path can be followed in Figure 4.2. Thus, at an income of Y_2 per annum, an individual's demand for, say, sausagemeat is at a maximum — at, say, Q_2 per annum. If income rises above Y_2 to Y_3, then the individual can afford to buy better-quality meat and therefore substitutes lean meat for the sausages. As a result, someone earning Y_3 may consume the same quantity of an inferior product, Q_1, as someone on an income of only Y_1. With normal goods, the greater the income, the more will be demanded, although as income rises the rate of increase in consumption tends to slow down.

The quantity demanded is not only influenced by the level of income but also by its distribution in a country. The distribution of income is measured by the Lorenz curve and Gini ratio. The Lorenz curve plots the cumulative dis-

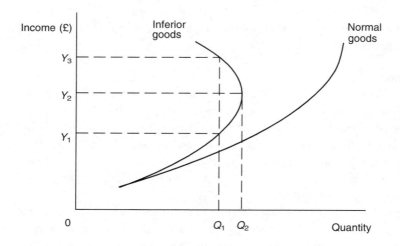

Figure 4.2   Engel curves of normal and inferior products

tribution of income as income rises. Population is measured along the horizontal axis and the cumulative percentage of income up the vertical axis. Thus at Point *A* in Figure 4.3, a third of the population earns around 10 per cent of income. At Point *B*, 90 per cent of the population earns just over half. Such a Lorenz curve represents a more unequal distribution compared to the dotted Lorenz curve. The more equal the distribution, the nearer the Lorenz curve shifts towards the 45-degree line. This visual interpretation of the distribution of income can be given mathematic precision. The ratio of the area of the segment of the curve to the area of the triangle is known as the Gini ratio. The lower the Gini ratio, the more equal the distribution of income.

If income is unequally distributed, effective demand will tend to come from the relatively few people with sufficient funds to place orders. This does not mean to say that the rest of the population has no demand or desire, but their willingness to purchase is not backed by money and hence is ineffective in distributing goods in their direction. Given the same GNP, the more equal the distribution of income, the greater the demand, as more people would have income with which to make their purchases. Tax policies aimed at redistributing income raise demand.

There is a second reason for an increase in demand arising out of such 'Robin Hood' tax policies. Poorer people tend to spend a larger proportion of their income than do richer people, and wealthier people tend to save a higher proportion of their income than do poorer groups. As income is

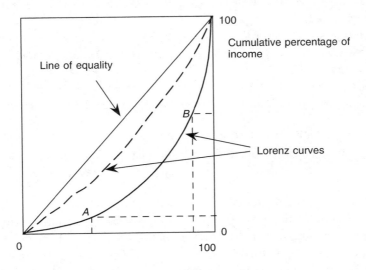

Figure 4.3   The Lorenz curve

redistributed, a higher proportion of it will be spent. This will increase the effective demand of the least well-off sections of society.

Size of the population

The quantity demanded tends to vary directly with the size of the population. Because the expected life of buildings may often exceed fifty years, it is important for the long-term viability of projects to take possible population changes into account, even when the developer expects to sell a property on completion. The value of the property at any point in time contains assumptions about demand arising from the size of the local population well into the future. Changes in population size are determined by the relationship between the birth rate, which is the number of births per 1000 members of the population, and the death rate, which is the number of deaths per 1000. The change in the size of the population brought about by the number of births and deaths needs to be modified by the level of net migration, which is the difference between the numbers of people immigrating and emigrating. This produces a rate of growth of population.

For example, the population will expand when the birth rate exceeds the death rate, unless net migration is so great that it reduces the total population; if far more people leave a country or locality than enter it. In the period from 1972 to 1982 the population of the UK declined by almost 430 000: more people emigrated than immigrated. While 227 400 left these shores each year on average, only 184 000 came in to replace them. This decline in population occurred in spite of a drop in the death rate, which raised the numbers of older people. Longevity, or life expectancy, rose to 70 years for males and 76 for females.

A useful rule of thumb to establish the length of time needed for a population to double in size is to divide 70 by the percentage rate of growth:

$$\text{The percentage rate of growth} = \frac{\text{Change in population in a given year}}{\text{Population at start of year}} \times \frac{100}{1}. \tag{4.2}$$

Thus, if the growth rate for an area or city is 2 per cent per annum, one might expect the population to double in thirty-five years, if the rate is sustained. Such growth rates are by no means unusual and they imply large increases in population, well within the life expectancy of most buildings. Therefore there may often be a need to take longer-term population trends into account during the planning stages, to protect the long-term viability of projects.

Various distributions of population can be used to assess potential needs. Population can be analysed in terms of income, age, employment, education and most importantly household formation, in order to establish the kinds of building required. For example, new schools may be needed for an expanding young population, or extra sheltered housing for the growing number of elderly people. Household formation refers to the rate at which young people leave home; divorce rates divide households into two separate households; and lifestyles determine the size of households. Households may consist of families, flatmates sharing, or single people living alone.

Long-term building trends reflect the growth in the size of the population. Periods of rapid population growth in the nineteenth century were therefore also periods of greater levels of building activity compared to the 1840s and 1880s or even the 1900s and 1910s, when both population and housing stock grew more slowly. Building trends also reflect the way people lead their lives: although the total overall population of the UK became smaller, the number of households increased from 18.3m in 1971 to 20m ten years later. The average number of people in each household dropped from 2.9 to 2.6. At the same time the proportion of all households comprised of only one person rose from 18 to 23 per cent.

Taste

Tastes or individual human preferences reflect culture, fashion, habit and even climatic conditions. Society's tastes can be influenced by advertising and other promotional efforts. Taste will, in turn, influence the demand from individuals as well as companies for particular styles or types of building. Corporate clients may well have a desire to achieve a certain company image. The recent growth in demand for one-bedroom properties in London may be attributed to a change in tastes and fashion as well as lifestyles. The desire to live in houses has led many people to move from cities to more rural locations where land and property prices are lower and which offers a spacious, quieter lifestyle than living in a city.

The location of a building is also an important determinant of demand. Proximity to markets may be a major consideration for firms, while living in a pleasant neighbourhood, perhaps near to schools, shopping facilities and public transport would often be a high priority for house buyers. As far as this economic analysis is concerned, these location factors form part of the background to buyers' tastes, and hence to decisions to purchase particular properties. A change in taste in favour of a product will raise demand, assuming everything else including price remains the same. A change in taste away from the product will reduce the quantity demanded.

Government policy

Government policy is another major determinant of demand, sometimes of overriding influence in the building industry, since such a large proportion of demand comes from government sources. The government may, for example, decide to increase the provision of public sector housing as a matter of policy. In 1919, Christopher Addison's Housing and Town Planning Act required local authorities to provide adequate council housing. Again in the middle of the century, as Lewis (1965) points out, it was official policy to complete 175 000 houses a year in England and Wales. In the 1980s, legislation obliged councils to sell their properties to sitting tenants. At the same time, local authority house building programmes were cut back as a result of central government policy and rate capping, which penalised those local authorities that continued to spend beyond limits set by the Department of the Environment in England and Wales and by the Scottish Office north of the border.

Expectations

Expectations of price changes will affect the quantity demanded at any one time. For example, the demand for housing may rise if the population anticipates that house prices will rise. This increase in demand will often become part of a self-fulfilling prophecy, as shortages in housing may then lead to higher prices being asked. If house prices are expected to fall, then many potential buyers will be deterred from entering the housing market, and demand for housing will decline.

These expectations of future price changes have to be distinguished from current prices. As we noted above, the higher the current price, the less will be demanded, assuming everything else remains the same. Yet here we see that expectations of higher prices in fact raises demand. This is because people realise that if they buy at the current price, they will be able to sell at a higher price at some time in the future, if their expectations are correct. The lower the price, the more will be demanded, but if people expect prices to fall even further they may be put off making a purchase regardless of the current price. Nevertheless, given the same expectations, the lower the price, the greater will be demand.

Shifts of the demand curve

In Figure 4.4, when a price changes there is a movement *along* the demand curve, either D_1–D_1 or D_2–D_2. When any factor other than price changes, the conditions in the market alter. Depending on the type of market and the nature of the change, the demand curve will either shift outwards or inwards. The changed circumstances may lead to more or less being demanded *at the same price* as before. Thus, at price P_2, the quantity demanded increases from Q_1 to Q_2, or decreases from Q_2 to Q_1, depending on the nature of the change in circumstances. Alternatively, the curve may shift upwards or downwards, indicating that consumers demand the same quantity as before but are now willing to pay a higher or lower price. Thus, for example, as a result of a down-ward shift in demand, consumers or buyers would be willing to pay only P_1 instead of P_2 for the same quantity of Q_2 units.

When firms such as Coca Cola and Pepsi Cola advertise, they do so in order to encourage people to buy more of their respective products at the prevailing price. In terms of Figure 4.4 they attempt to shift demand horizon-tally. In the property market, however, a shift in demand as a result of a change in interest rates, for example, means that buyers are willing to pay higher or lower prices for an *existing* dwelling, office or premises. This may be interpreted as a vertical shift in the quantity demanded.

Elasticity of demand

If an estate consisting of 2- and 3-bedroom houses is built and sales are slow, the question arises, should prices be reduced? And if sales are rapid, should

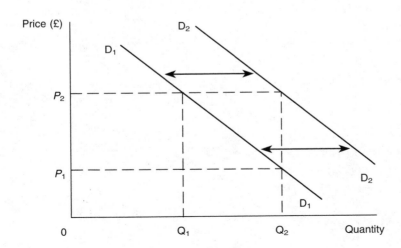

Figure 4.4   Shifts of the demand curve

the prices of remaining houses be raised? If a professional practice of architects or quantity surveyors is very busy, should it increase its fees? Part of the advice given in such circumstances must look at the consequences of price changes. Elasticity of demand measures the responsiveness of demand to a change in something else. Price elasticity of demand, therefore, is a measure of the responsiveness of the quantity demanded to changes in price. For example, if price rises, consumers will buy less, but how much less? The drop in sales affects revenues as well as the amount of work the firm takes on, and therefore the number of people it needs to employ. It follows that there is a need to know the extent of the fall in demand in order to plan accordingly.

From the first law of supply and demand, the lower the price, the more will be demanded. Demand for the services of a practice may increase if it reduces its fees, and manages to undercut its competition. The question remains, however, will the total revenue coming into the practice increase or decrease? Will the increase in demand be sufficient to compensate for the reduction in price?

To find the value of the price elasticity of demand, use the following formula:

Elasticity of demand

$$= \frac{\text{Proportionate change in quantity demanded}}{\text{Proportionate change in price}} \qquad (4.3)$$

For example, if a price is raised from £10 to £11 and the quantity demanded falls from 50 to 40 units, then the price elasticity of demand can be calculated as follows:

$$\text{Elasticity of demand} = \frac{\dfrac{q_1 - q_2}{q_1}}{\dfrac{p_1 - p_2}{p_1}} \qquad (4.4)$$

where q_1 = quantity before change
        q_2 = quantity after change
        p_1 = price before change
and p_2 = price after change

Let q_1 = 50 units
        q_2 = 40 units
        p_1 = £10
and p_2 = £11

Substituting in Equation (4.4):

$$\text{Elasticity of demand} = \frac{\dfrac{50 - 40}{50}}{\dfrac{10 - 11}{10}} = \frac{\dfrac{10}{50}}{-\dfrac{1}{10}} = \frac{10}{50} \times \frac{10}{-1} = \frac{100}{-50} = -2$$

The value of price elasticity of demand will always be negative, reflecting the negative downward slope of the demand curve. However, for the purpose of interpreting the result, the minus sign may be ignored. In mathematical terms, we are interested in the modulus, which is the value of the number alone, ignoring whether it is positive or negative. In this case, the price elasticity of demand is equal to -2 (a modulus of 2). The range of values of elasticity of demand lies between infinity and 0. If the calculation produces a modulus greater than 1, then the price elasticity of demand is said to be elastic. If it is less than one elasticity of demand it is said to be inelastic. In this instance, the modulus is less than 1. A rise in price would be accompanied by only a relatively small reduction in demand. Consequently, a rise in price would also raise total revenues, because the increase in the quantity demanded is more than sufficient to compensate for the reduction in sales.

Table 4.1 summarises the effect of proposed price changes on the quantity demanded and total revenue. Total revenue is the price or average fee multiplied by the number of projects. Thus, if the responsiveness of demand is elastic, a price increase will reduce the quantity demanded so dramatically that revenues will decline in spite of the price rise. However, if the elasticity of demand were inelastic, the decline in demand would be slight. Although some clients may go elsewhere, sufficient numbers would remain so that the higher price charged would more than offset the loss of custom.

When the proportionate change in price is equal to the proportionate change in the quantity demanded, the elasticity of demand is equal to 1, or unity, resulting in no change in total revenue when price is altered. However, a price rise under these circumstances would result in a drop in the quantity demanded, even though revenues remained unchanged. Thus the same revenue may be achieved with less effort, simply by raising charges. If demand is inelastic, then a price increase would reduce demand only slightly, and total revenue would rise. In such circumstances, it would be possible to raise fees, thereby reducing workload while increasing income.

Before taking any action to raise or lower a price, there are, of course, many other factors which need to be taken into account, such as the degree of competition, relations with clients and the kind of projects that may be lost or gained by any proposed price changes. Above all, the effect on profitability has to be considered. It is worth noting that increasing total revenues does

Table 4.1   The effect of price changes on the quantity demanded and revenues

| Price elasticity of demand | | Price change | Quantity change | Revenue change |
Value	Description	(Cause)	(Effect)	(Effect)
>1	Elastic demand	Up	Down	Down
>1	Elastic demand	Down	Up	Up
<1	Inelastic demand	Up	Down	Up
<1	Inelastic demand	Down	Up	Down
=1	Unit elasticity	Up	Down	Same
=1	Unit elasticity	Down	Up	Same
=∞	Perfectly elastic	Up	No demand	Zero
=∞	Perfectly elastic	Down	As much as is produced	Up
=0	Perfectly inelastic	Up	No change	Up
=0	Perfectly inelastic	Down	No change	Down

not necessarily increase profits. Changes in demand may lead to changes in production plans, which will affect costs. Increasing demand and total revenue could turn out to be counter-productive, if costs rise even faster than revenues. It is quite possible for total revenues to rise, but the increased demand may call for increased overtime payments which are greater than the increase in revenue. Reducing the quantity demanded may reduce revenue, but it may also reduce costs by an even larger margin. The result would then be an increase in profits, at least in the short run.

Three examples of price elasticity of demand are illustrated in Figure 4.5. The vertical line (price elasticity of demand = 0), represents a perfectly inelastic demand curve, since the quantity demanded remains the same regardless of any price changes there might be. The horizontal demand curve (price elasticity of demand = ∞), is perfectly elastic, since an infinitely small increase in price will result in absolutely no demand for the product or service. The curve (price elasticity of demand = 1), a rectangular hyperbola, represents a demand curve with unit elasticity. These are the only examples of demand curves with uniform values. In all other cases the price elasticity of demand varies as one moves along the demand curve, even when they are depicted as straight lines, as in Figure 4.4.

We have seen that price elasticity of demand measures the responsiveness of the quantity demanded to changes in price. Other variables which may cause demand to change are changes in income and changes in the price of some other product or service. Income elasticity of demand measures the responsiveness of the quantity demanded to changes in income. Cross elasticity of demand measures the responsiveness of demand to a change in the price of something else. The more competitive two practices are with each

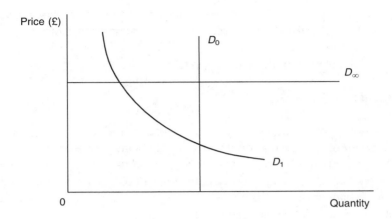

Figure 4.5   Price elasticity of demand: unit elasticity, perfectly elastic and perfectly inelastic demand curves

other, the higher the cross elasticity of demand between them. These elasticities may be calculated as follows:

Income elasticity of demand
$$= \frac{\text{Proportionate change in quantity demanded}}{\text{Proportionate change in income}} \qquad (4.5)$$

Cross elasticity of demand
$$= \frac{\text{Proportionate change in quantity of A demanded}}{\text{Proportionate change in price of B}} \qquad (4.6)$$

Supply

It is, of course, extremely difficult to quantify the service an architect, quantity surveyor, lawyer, or any professional practice, supplies. Nevertheless, the supply of a professional service may be indicated by the amount of time devoted to particular projects by a number of architects, though this does not take the quality of the service into account, nor the ability of the architects. Supply is the quantity of the service architects are willing and able to offer clients over a given period of time. The quantity supplied may refer to individual practices or to the profession as a whole. Like the quantity demanded, the quantity supplied also depends on the combined effect of a variety of factors, including the fees charged, the cost of premises, labour costs, the price of other services provided by the practice, the state of technology and the aims of the practice. We shall look at each of these factors in turn.

Price and fees

The quantity supplied depends on the price of the product or service. Hence, the quantity of the service supplied by, say, architects is determined by the fees they can charge. The second law of supply and demand applies assuming that everything else remains constant, the higher the fee, the greater quantity of time architects are in a position to supply; the lower the fee, the fewer the number of hours that will be spent on a given project.

Figure 4.6 illustrates a supply curve to show the relationship between, say, the number of hours an architect may be willing and able to devote to a project, and the fees charged. In return for a fee of £P_1, Q_1 hours would be committed to a project. If the fees had been £P_2, then Q_2 hours could have been provided. Of course, in practice, designs cannot be left half completed just because insufficient time has been allowed. The supply curve here only illustrates the number of hours a practice would plan to spend on designing a project for a given fee. Many other factors also have to be taken into account.

Supply curves may represent individual practices, or the profession as a whole. In the latter case, the individual supply curves of all practices would simply be aggregated to form the industry or profession's total curve.

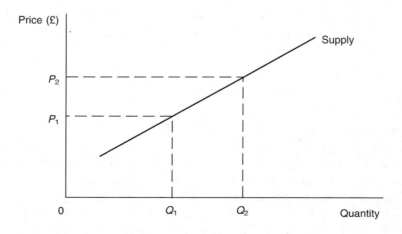

Figure 4.6   The supply curve

Shifts of supply

Figure 4.6 shows that a movement along the supply curve occurs if price changes but everything else remains the same. However, if something other than price changes, then the whole supply curve will shift outwards or inwards. Figure 4.7 illustrates that, depending on the nature of the change of circumstances, at every price, either more or less will be supplied. Thus, at price £P_2 the quantity firms are willing to supply would change from Q_1 to Q_2, or from Q_2 to Q_1, depending on the cause. Similarly, if the quantity remains the same, at say, Q_1 units, then the price charged would change from P_1 to P_2, or from P_2 to P_1. For example a firm might introduce a new technique into the building process, which has the effect of shifting the supply curve from S_1 to S_2. As the firm wishes to win more orders, but it cannot increase the amount of work available, it may decide to carry out the same amount of work at a reduced price. The new technology enables the firm to reduce its price without increasing its output in order to remain competitive.

Products in competitive supply

Other factors influencing supply in a particular period are the amount, price and profitability of alternative work. We have seen that the higher the price, the more will be supplied. When the price of work on alternative sites goes up,

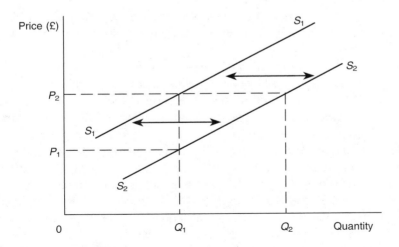

Figure 4.7   Shifts of the supply curve

firms will concentrate their efforts there rather than the current site on which they may be working. As a result, even though the price and work on a particular site might not have changed, the willingness of firms to work on a site can go down. In practice, this means that subcontractors may start a job and then allocate labour to another, more profitable, site. It is as if sites were competing for the attention of subcontractors, and the supply of subcontractors depends on the price offered at alternative sites. In terms of Figure 4.7 the supply line shifts to the left from S_2 to S_1 when alternative sites offering higher wages and profits attract workers away from a particular job. At $£P_2$, the supply of labour goes down from Q_2 to Q_1 workers.

The factors of production

Production of any sort relies on various inputs, just as baking a cake requires labour, equipment and ingredients. These inputs are called the *factors of production*, and in neoclassical economics include land, labour and capital. The price of land is rent, the price of labour is wages, and the price of capital is interest. Enterprise is also seen as a factor of production but it is different from the others. It has no price, because an enterprising individual receives profit only after all other costs have been met.

Land, as a factor of production, is where business is conducted. For most firms it consists of the premises in which work takes place. However, land as a factor of production also includes the minerals it contains. Thus, quarrying for aggregates to be used in buildings or works involves land as a source of raw materials. Whereas in most industries production takes place in buildings owned or rented by the firm, in construction, the production process takes place on land owned by the client,

Labour, as a factor of production, refers to the human input which takes time, such as labouring on site, or time spent at a desk in an architect's office. It embodies the skills and knowledge of the people employed and therefore the quality of labour will depend on the training, ability and motivation of the individuals employed.

The third factor of production is *capital*, which is the total value of the assets of a firm. Capital refers to the equipment, buildings and materials owned by a firm, as well as the working capital or money needed to finance work until payment is made by clients. Firms may also hold assets in the form of retained profits they have saved from previous years and which they have invested, for example, in the shares of other firms. Because capital is a factor of production, it determines the size of the firm. The management of capital assets is a major determinant of a firm's profits.

A firm may own or rent its premises. If it owns them, the money invested in the building does not earn interest but instead saves the firm from paying rent. If the rent the firm would have paid for the building is greater than the interest lost, then the firm is better off than it would otherwise be. Any spending on assets must be compared to the highest return the firm would have earned in an alternative. Economists call this comparison the *opportunity cost* of an investment. Bank interest rates are a useful comparison, since they represent the rate of return on money with minimum risk. The point is that when interest rates rise, many firms will feel that the extra risk of investment is no longer worth taking. If interest rates rise, the quantity supplied will be reduced.

Similarly, the price of all or any factor of production will affect the quantity firms can supply. If, for example, salaries paid to employees rise, the ability of a firm to provide a given quantity of time and effort at a given price may be adversely affected. Projects may no longer be financially viable or profitable. Conversely, if salaries or rents go down, then projects which might not have been viable previously may become profitable and attract increased competition or supply. Again changes in the cost of factors of production cause the supply curve to shift to the left or right. Assuming the selling price of the firm's output remains the same, either more or less can be supplied.

Aims and objectives

We have noted above that the size of a firm is partly determined by the amount of its capital. In order to survive, the value of a firm's capital assets must rise at least in line with its competitors, otherwise, the firm runs the risk of being taken over by one of its rivals. To survive, each firm must make profits, part of which it can plough back into the firm in order to expand. Firms need profits for investment in themselves.

In neoclassical economics, one of the main aims of firms is assumed to be to maximise profits; that is, to ensure that profits are as high as possible. Firms may aim to maximise their sales or increase their share of the market. However, in order to increase sales, costs may rise or fees may have to be reduced and actual profits may decline. Often, therefore, the business aim of firms is to maximise sales within a profit constraint. Sales are not encouraged beyond a certain level, if, as a result, profits would drop below a target figure. Other aims include maximising returns on investment capital and increasing the value of shareholders assets.

The aims and objectives of professional practices are also determined by the preferences of the partners. This is an important point, because most design and professional practices in construction are partnerships, owned and run by a group of people who work within the firm. There are few outside shareholders with power and influence over the running of the firm. The partners, because of their own expertise, experience or interests, may wish to concentrate on a specialist service in a particular area or building type. The firm would thus gain a reputation for certain specialist skills. Each practice needs to distinguish itself from its competitors, since the perception of a practice by potential clients is vital for marketing its services. Economists call this marketing practice *product or service differentiation*.

Thus aims and personal preferences have to be included as one of the factors influencing the quantity firms are willing to supply. If the aims or strategy of firms change, supply will shift accordingly, because firms will raise or lower output depending on their aims, even without a change in price.

Technology

The use of new technology increases the capacity of firms to produce: firms can either produce more at the same price or the same output at a lower price. This can be seen in Figure 4.7. Before the introduction of new plant and equipment, the supply curve is given as S_1. At price £P_2 firms offer Q_1 units for sale. The introduction of new techniques shifts the supply curve to the right to S_2. At every price, more could be supplied. Firms can therefore now produce Q_2 units at the same price as before, or they can still produce Q_1 units at the lower price of £P_1.

Computer aided design (CAD), computer aided management (CAM), and continuous improvements in the plant and machinery used on site in theory therefore increase the potential quantity or capacity of the building industry. This means that more could be built for the same cost, or similar buildings could be built for less. Although innovation or the spreading of new techniques and technology has the potential to reduce construction costs, in practice these gains are rarely achieved in the building industry. In the previous chapter, it was seen that any reduction in construction costs were not passed on to buyers but instead enabled developers to offer increased bids to landowners in order to purchase sites. The beneficiaries of lower construction costs are not end users, but landowners. Final building costs remain the same and there is no increase in demand as a consequence of reduced building costs lowering building prices and therefore increasing the demand for buildings. As a result, the main point of innovation in construction is to reduce costs in order to remain competitive with other firms in a market, the size of which suppliers cannot influence by reducing their prices. In other industries, new technology leads to lower prices and increased demand and output. New technology in construction is not used to increase overall demand for construction output, or, for that matter, to increase the aggregate capacity of construction firms. In fact, contractors tend to hire equipment as and when required.

It is likely that those firms which adopt the new technology fastest and most effectively will be those that remain the most competitive. The adoption of new technology and new techniques enables construction firms and professional practices to remain competitive internationally. There is also a need to remain competitive in the home market as architects from all over the world — and especially from other countries in the European Union — continually adopt and upgrade new technology and offer their services in the UK market.

Exogenous factors, government policies, and nature

Exogenous factors are those aspects of society, and even nature, which lie outside the scope of economics but which have an impact on economic variables, such as the quantity supplied. For example, wars and political changes can affect an industry's ability to supply if communications and transport links are broken. Political uncertainty in some countries deters contractors from entering certain markets. Accidents and strikes can have important economic implications on particular sites or for particular firms. Nature, which includes weather conditions and the geological difficulties of a site, may also affect the supply of construction services.

Government policies can either encourage or discourage production. Grants encourage building and works in certain regions, or inner cities. Elsewhere, planning constraints and public objections to motorways and out-of-town shopping centres reduce demand for construction. Interest-rate policy, and decisions to invest in infrastructure projects, have direct implications for construction work, in particular because the public sector remains a major component of construction demand.

Finally, as production decisions are taken in advance of production, and production takes place in advance of sales, expectations play an important role on the supply side, just as they did on demand. Expectations of future developments, new products and new markets at home and abroad mean that firms must gear up in order to prepare for future demand. Construction, however, remains a very uncertain industry for contractors, who do not know which of them will win contracts to build. Often they do not know until shortly before going on site what materials, methods and skills will be required. As a result, very little planning is possible in construction compared to other industries.

Price elasticity of supply

The construction industry responds to demand. It is only after orders are placed with contractors that construction takes place. The question's are, how responsive is construction? What factors have an impact on construction output? How long does it take for individual firms to respond to changes in price? To answer these questions, elasticity of supply measures the responsiveness of the quantity supplied to changes in the economy. Price elasticity of supply, for example, measures the responsiveness of the quantity supplied to changes in price.

Just as price elasticity of demand relates changes in the quantity demanded to changes in price, so price elasticity of supply relates changes in price to the quantity offered for sale. The formula for price elasticity of supply is:

$$\text{Price elasticity of supply} = \frac{\text{Proportionate change in quantity supplied}}{\text{Proportionate change in price}} \tag{4.7}$$

For example, in a local housing market, if the average price of a 3-bedroom house rises from £100 000 to £110 000, and the number of such houses available in estate agents increases from 100 to 115 houses, then the price elasticity of supply can be calculated as follows:

$$\text{Elasticity of supply} = \frac{\dfrac{q_1 - q_2}{q_1}}{\dfrac{p_1 - p_2}{p_1}} \tag{4.8}$$

where q_1 = quantity supplied before change
         q_2 = quantity after change
         p_1 = price before change
and p_2 = price after change.

Let q_1 = 100 houses
         q_2 = 115 houses
         p_1 = £100 000
and p_2 = £110 000

Substituting in Equation (4.8),

$$\text{Elasticity of supply} = \dfrac{\dfrac{100-115}{100}}{\dfrac{100\,000-110\,000}{100\,000}} = \dfrac{\dfrac{-15}{100}}{\dfrac{-10\,000}{100\,000}}$$

$$= \dfrac{-15}{100} \cdot \dfrac{100}{-10} = \dfrac{15}{10} = 1.5$$

When elasticity of supply is greater than 1, supply is said to be elastic. For example, if the price elasticity of supply of housing in a given locality is 1.5, then a 10 per cent rise in price will be accompanied by an increase in supply of 15 per cent. If supply were inelastic, say 0.5, then the same 10 per cent increase in price would only raise supply by 5 per cent. Notice that the price elasticity of supply is always positive, reflecting the positive slope of the supply curve. Price elasticity of supply is illustrated in Figure 4.8. Perfectly elastic supply is horizontal, while inelasticity of supply is shown as a vertical line. Unit elasticity applies to any straight line supply curve starting at the origin.

The concept of price elasticity of supply helps to predict the behaviour of firms in response to price changes. Governments use this approach to anticipate demand for imports, tax revenues and employment in the construction industry. However, it is far more difficult to measure and predict the output of firms than to do the same for the supply of housing in a local housing market. It is especially difficult to predict the elasticity of construction supply. In order to plan an increase in the supply of construction services there is a need to know the number of skilled people available in firms and those who would be attracted into working in construction if required. The period of training needed would cause delay if skill shortages meant that people were not available. The delay could be shortened by using unskilled and inexperienced labour at the expense of quality. These considerations make measuring the response of construction difficult.

Nevertheless, the ability of firms to respond to price changes determines the elasticity of supply, and this will vary with time. In the very short term, the elasticity of supply may be perfectly inelastic, meaning that for any price change there is no change in the quantity coming on to the market. Reducing output when the production of goods has already commenced can be as problematic as increasing output at short notice. In most industries, the longer firms are given to respond to changing circumstances, the greater the elasticity of supply.

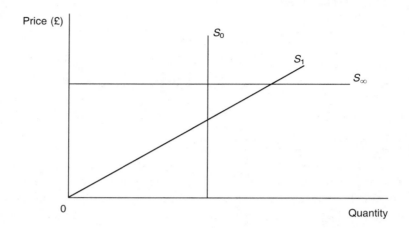

Figure 4.8   Price elasticity of supply: unit elasticity, perfectly elastic and perfectly inelastic supply curves

In construction, there is a paradox. As Ball (1988) has pointed out, although individual firms have to be flexible in order to respond to great variations in demand for their services, the construction industry as a whole is less flexible than most sectors of the economy. Firms tend to hire from pools of resources. They hire plant when needed and take on labour only for the duration of projects, or even work packages within projects. The process of developing a site is also complex and requires time to put into action. In order to expand output, developers must first obtain a site, gain planning permission, design the building and invite tenders from contractors.

In the property market, once a building has been completed, it remains in existence for many years, even when demand may have dropped and it becomes surplus to requirements. The property might then remain empty, reducing the utilisation of the stock of buildings. Again, in other industries, a surplus would lead to lower prices. In the construction industry, prices tend to be 'sticky' downwards. This implies that rents will tend to rise or remain the same rather than go down. Rental agreements or leases are usually longer-term arrangements, based on the current market situation at the time the contract is signed, with periodic upward-only rent reviews.

Elasticity of supply therefore tends to be inelastic in the short run. The quantity of buildings coming on to the market at any one time is unresponsive to changes in market conditions, at least in the short run. As a result, when shortages of property do occur, prices and rents can soar. During the mid-1980s house prices in the South East rose at the rate of between 25 and

30 per cent per annum, depending on location, only to fall by similar percentages in the early 1990s.

Having discussed the economic influences of supply and demand, we now turn to the broader macroeconomic influences of inflation and unemployment to see how they influence the economics of the construction industry.

Inflation and the use of index numbers

Inflation is the rate at which prices rise on average. The rate of inflation is usually given on an annual basis, but monthly rates are also produced. It is a major economic factor affecting the construction industry, adding uncertainty to budget forecasts. Costs can rise more than expected and affect cash flows adversely. Of course, property prices themselves respond to inflationary pressures, which means that if it is anticipated, inflation in costs during the construction phase can be absorbed in the form of higher property selling prices.

Broadly speaking there are two types of inflation, distinguished from each other by their causes rather than their effects. *Cost push inflation* occurs when rising costs are passed on to customers as firms raise their prices to maintain their profit margins. On the other hand, *demand pull inflation* is caused by excess demand in the economy bidding prices upwards. Aggregate demand consists of demand for consumer goods and services, such as cars, food and haircuts, investment demand for plant and machinery, and government spending on roads, hospitals and schools. When aggregate demand is greater than the total of goods and services firms can supply, there is excess demand in the economy.

The causes of inflation remain open to interpretation. Cost push inflation in one period may be the result of pressures arising out of a period of demand pull inflation in the previous period, which in turn might have been preceded by a cost push inflation, and so on through a spiral of rising prices.

The rate of inflation is measured each month by the Office for National Statistics by calculating the annual rate at which consumer prices rise, using the retail price index (RPI). The RPI is based on the value of a weighted average basket of consumer goods and services, including housing costs. The weights refer to the quantities of the various items to be included. As these weights reflect consumer spending patterns, they are updated periodically to take new products into account as well as changes in patterns of spending. Each unit price is multiplied by its respective weighting. The weighted results are then aggregated to give a total value. This is then compared to the base year total to give a percentage change. This produces the index number. The base year always equals 100. If the increase between the base year and the current year is, say, 36 per cent, the index number is given as 136.

Thus, if the total value of a weighted average basket of goods was £300 in the base year, it would be given the value 100. If the value of the same basket of goods rose to £330 in the following year, then the new index number is calculated as follows:

New index number

$$= \frac{\text{New total value of goods and services}}{\text{Original value of goods and services}} \times \frac{100}{1} \tag{4.9}$$

Hence,

$$\text{New index number} = \frac{£330}{£300} \times \frac{100}{1}$$
$$= 110$$

The annual inflation rate is therefore 10 per cent. The rate of inflation is based on an annual moving average, giving the yearly rate of average price increases. An annual moving average is calculated every month. Each month the first month in the previous annual calculation is dropped. The rise in the index over the previous month is then added to the preceding 11 months. The price index is therefore always calculated on the previous twelve months' figures.

The rate of inflation reached a peak in 1975 of 24 per cent, falling to 8 per cent in 1978, and rising again to 18 per cent in 1980. In the 1980s the rate of inflation dropped and by October 1987 it stood at 4.5 per cent, although it rose again towards the end of the decade. During the first half of the 1990s the average rate of price increase as measured by the RPI has been around 3 per cent per annum.

The RPI reflects price movements, usually upwards, facing the average consumer. All goods and services are taken into account, but not all people buy all goods: if the price of, say, cigarettes goes up, the rise will only affect smokers. Different categories of people and households buy different baskets of goods. The relative weights of different goods varies from person to person. Therefore, different individuals face different rates of inflation in their cost of living. The RPI is only an indicator of the general movement in consumer prices.

The headline rate of inflation is the RPI, while the underlying rate of inflation is measured by the RPI, excluding housing costs. Other price indices include the producer price index. The producer price index measures the rise in materials and labour costs facing firms, while the pensioner price index is identical to the RPI apart from the exclusion of housing costs. Similarly, construction cost and price indices take into account the movement of prices affecting construction. These specialised construction industry indices are necessary because the RPI refers to consumer goods and services and may not reflect price increases facing clients, builders, quantity surveyors and architects.

If building prices rise at the same rate as the rate of inflation, the real cost of building does not change in terms of the average basket of goods. However, building costs have been rising faster than the rate of inflation for most of the past hundred years. The real cost of construction has risen. The building cost index (BCI) and the tender price index (TPI) are used to measure the rate at which building costs and tender prices rise per annum. Several other price indices are used in the construction industry, including the mechanical cost index and the electrical cost index.

The BCI includes movements in the prices of basic material, and agreed wage rates and contractors' on-costs. This cost index can be used to anticipate a rise in prices in the period between the date a tender is submitted and the completion of a project, especially on large projects and those taking several months or years to construct. For this reason, price index tables often include several quarterly projections of building costs.

To estimate the final building costs of a project, first find the total cost of a project from suppliers and subcontractors. This figure then forms the current total cost used as the basis for the tender price or bid price for work. The expected final building cost is calculated using the following formula:

Expected final building cost

$$= \text{total current building costs}_t \times \frac{\text{BCI}_{t+1}}{\text{BCI}_t} \qquad (4.10)$$

where,

Total current building costs$_t$ = total current building costs at tender date

BCI$_t$ = building cost index at tender date

BCI$_{t+1}$ = building cost index at expected completion date.

An example using a price index to adjust prices can be seen in Chapter 12.

An interesting picture emerges from a comparison of the indices shown in Figure 4.9. Firstly, average earnings of construction workers rose at a faster rate than construction material costs. This is a continuation of the long run trend of wages compared to material costs and accounts for the steady substitution of increasingly expensive labour relative to off-site prefabrication. Also, from Figure 4.9 the building cost index of all new construction output declined from 1989 ahead of the decline in tender prices for all public sector building contracts. Indeed the tender price index continued to climb until 1990. Only then did building costs drop sufficiently for contractors to reduce their prices also the lower building costs reflected the drop in the volume of work after 1990 and therefore the increased competition for work forced contractors to reduce their tender prices in their bid to win work.

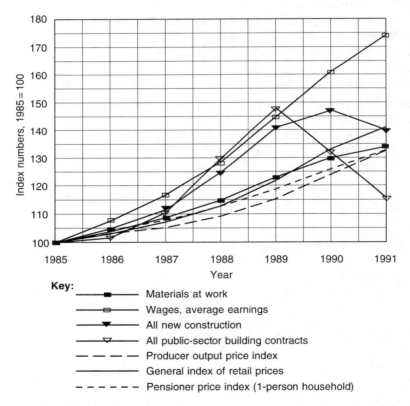

Key:

——■——	Materials at work
——□——	Wages, average earnings
——▼——	All new construction
——▽——	All public-sector building contracts
– – – –	Producer output price index
————	General index of retail prices
– – – – –	Pensioner price index (1-person household)

Source: Housing and Construction Statistics, 1992, table A; *Economic Trends,* 1993, table 26.

Figure 4.9   Construction cost and price indices and producer and retail price indices, 1985–91

Comparing the all new construction output price index to the producer output price index it can be seen that construction prices rose more quickly than manufacturers prices until 1989. Thereafter, while construction costs declined, producer output prices continued to rise, overtaking the construction output price index before 1991. The building cost index also rose at a faster rate than the RPI until 1989, but then building costs fell, allowing the retail price index to catch up. The retail price index grew more quickly than the pensioner price index, showing that housing costs rose at a rate above the average rate of price increase.

Although inflation may stimulate investment because investors can make speculative gains, it may also have a dampening effect on investment decisions, because of the uncertainty it creates regarding future prices. The impact of inflation is as much psychological, affecting business confidence, as it is economic. As long as inflation persists, there is always the threat that a rising rate of inflation will accelerate out of control. Such a situation is known as *hyperinflation*. Hyperinflation arises when money loses its value so rapidly that people become increasingly unwilling to accept the currency in payment of debts.

5 Economic Structure of the Construction Industry

Introduction

The construction sector consists of different types of specialist firm working together on temporary sites to produce buildings and civil engineering works. Once they have fulfilled their contractual undertakings, each firm moves on to other work with different firms on new sites. Apart from the main contractors, architects, engineers and surveyors, very few firms remain for the duration of a project. Temporary organisations are formed to manage projects. On completion of their contracts, the firms making up these temporary teams also disperse.

Site labour tends to be hired on a casual basis. The hiring system works as follows. A worker is taken on to do a particular job, sometimes called a work package. On completion the worker is paid and must find more work on another site. If the employer has another site, it may be possible for the worker to move straight to another job for the same employer. Otherwise, a new employer must be found. In order to get more work from an employer, workers have an incentive to perform well. However, if another job at a higher rate of pay is offered by a firm on a different site, it is possible for workers to leave one site without notice and move to the higher-paid job.

There is no contract of employment, although firms usually hold back the first week's payment so that labour is paid one week in arrears. This gives the firm some leverage over the worker, but the worker is nevertheless seen as being self-employed. If the workmanship is not good enough, the worker can be dismissed. Often work that is unsatisfactory is discovered after the individual concerned is no longer working on the site. Both main contractors and subcontractors employ people on this casual basis. Because of all the uncertainties created by this method of employing labour to produce the built environment, it is hardly surprising that management control and progress on site rarely go according to plan.

Construction firms

One way of looking at the structure of the construction sector is to look at the number of firms involved in different aspects of the industry. Figure 5.1 shows the percentage of construction firms that are general builders, civil engineers and specialist trades, and the contribution of each category to the value of final construction output in 1991.

For example, although civil engineering firms, and building and civil engineering contractors, make up less than 5 per cent of the number of firms, their contribution to the value of output is over 25 per cent. In themselves, these figures mean little. The reason that a relatively few firms with a civil engineering capability make such a significant contribution to the output of construction is because they tend to be relatively large firms undertaking large projects, such as motorways, or complex projects, such as power stations.

Almost all firms in the construction industry are either general builders or classified under one specialist trade or another. A breakdown of the various

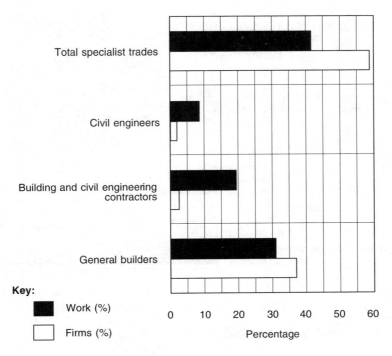

Source: *Housing and Construction Statistics,* 1992, tables 3.1 and 3.3.

Figure 5.1   Construction firms and output in 1991.

specialist trades, including firms such as plumbers and electrical contractors, can be seen in Figure 5.2. Many firms are, in fact, very small, often operating with only working proprietors and therefore employing only one person.

Of course, the number of firms is not the same as the number of people employed in each trade. To see how the number of people employed is related to the number of firms, Figure 5.2 shows the distribution of firms against the percentage of employees in each trade. To see how work is related to the number of people employed, Figure 5.2 also shows the relationship between the numbers employed in each trade and the value of the work done by that trade. Some interesting observations can be made from an analysis of the graph: for example, the value of the contribution of electrical contractors and heating and ventilating engineers together make up over 17 per cent of the

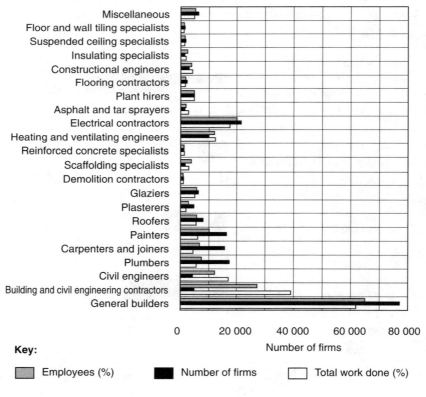

Source: Housing and Construction Statistics, 1992, tables 3.1, 3.3 and 3.14.

Figure 5.2   Distribution of firms, work, and employees according to trade, 1991

value of the output of all specialist contractors (that is, ignoring general builders, civil engineers, and building and civil engineering contractors).

We can see the data relating to trades and main trades, but it is necessary to understand that the ownership (and hence control of contractors and their relationship to their subcontractors) is often complex, involving holding companies and subsidiaries. For example, plant hire companies are often owned by large contracting firms which hire out plant and equipment to other contractors as well as making their own plant-hire divisions compete with outside plant-hire firms to ensure competitive pricing by the parent contracting firm.

The largest firms in the construction sector are usually owned by conglomerates, with interests in several industries. For example, firms such as P&O, and Trafalgar House own or have substantial interests in contractors. P&O own Trollope & Colls, while Trafalgar House own Bovis. Other firms, such as Tarmac, have several subsidiaries, involved in quarrying, contracting, civil engineering and plant hire. These complex financial relationships between contractors and other firms is significant for construction in several ways. First, it gives the major contractors a financial pedigree which strengthens their negotiating position with banks and other financial institutions as well as with subcontractors and materials suppliers. It enables them to take on large contracts at a moment's notice, since external funding is readily available to them. This makes them extremely flexible organisations, able to respond to changing circumstances in the market. At the same time, profits made in the construction sector by these companies are used to invest in other sectors of the economy. This reinforces the tendency for low levels of investment in construction, where the financial risks associated with research and development by contractors are high.

Although not included in Figures 5.1 and 5.2, the construction process also involves component and materials suppliers to the sector, some of whom produce almost entirely for construction purposes, such as brick manufacturers. Architects and other professional practices are also part of the construction sector, even though they are excluded from most statistical analyses of the industry. Before discussing construction markets, it may be useful to remind the reader of tendering and procurement systems, which describe the way building and civil engineering firms obtain work and relate to construction professionals.

Tendering

Many construction markets are characterised by the tendering process. The system operates as follows. When a client or firm wishes to hire a contractor or specialist firm, advertisements are placed and several firms will be approached. Each firm is given tender documents which include the design, specification and quantities of work required. The contractors quantity surveyors then price the bill of quantities and submit bids in sealed envelopes. Usually, the lowest tender price offered wins. There are two types of tendering process, open and selective tendering. In open tendering an advertisement might appear in trade journals, and any number of firms can apply to bid for the work. In selective tendering, the client prepares a short list of applicants, usually of no more than six firms. This increases the chance of any one tenderer being successful. In open tendering, the chance of success are reduced because of the number of competing firms. The high risk of failure deters many firms from participating.

Materials and labour costs are included in bills of quantities, which normally incorporate specifications as to how buildings are to be constructed, and to what quality. Buildings are analysed in terms of the quantities of work and their respective unit prices, and involve estimates of the labour and materials costs. The final total forms the basis for the tender. Bills of quantities allow comparisons to be made between different contractors tendering for a project, on the basis of the same work, starting dates and completion.

In practice, the actual cost of construction is different from the tender price, for a variety of reasons. Soil conditions may present unexpected difficulties. Adverse weather conditions often cause delays. Errors and accidents occur. Delays add to the cost of labour. Finance costs also rise if the client is unable to receive the revenues from the sale, rent or use of the building until the delayed completion date. Delays increase the period of outstanding loans, and hence interest payments.

Finally, there is a great deal of psychology and gamesmanship in the tendering process. Firms need to consider the degree of competition for a given contract. It is, for example, not a question of winning all tender contests the firm enters. If a firm finds it wins every time, it is possible it is underpricing the work. Either it could have charged a higher price and still won the bidding, or, alternatively, it might find it difficult to fulfil its contractual obligations. In tendering for work, if a firm has a full workload, it may not be necessary or possible to offer a low tender price. At the same time, it may be important for the firm to be seen to be interested, for future consideration for the same type of work.

Procurement methods

The structure of the construction sector can be examined in terms of the relationships and rivalries among the professions of the construction sector, mainly the architects and quantity surveyors, and between the professions and the trades. One purpose of studying the history of the construction sector in Chapter 2, especially over the last 150 years, was that the structure of the industry has emerged for historical reasons. It has been, and continues to be, an industrial battleground, which can only be understood in terms of the vested interests that have emerged as a result of past and present conflicts, technological developments and changing market conditions.

The battle for control over the construction process is fought over the choice of procurement method. As the construction sector is fragmented, the issue concerns the optimum method of having a building or civil engineering project built efficiently, to the quality specified, to cost and, of course, on time. The problems that every procurement method seeks to solve are, who is to lead the construction process, and who takes the risks. The procurement method allocates control to project leaders while spreading construction and financial risks on to other members of the building team and even back on to the client.

Although every building project is unique and every set of contractual relations reflects that fact, various common procurement methods are used to organise the production of a building or engineering project. In other words, though two separate projects may use the same general procurement method, rarely are the contractual arrangements identical. The four main types of procurement method are the traditional method of organising the construction process; construction management; management fee contracting; and design and build contracting. There are several variants of design and build, including design and manage, and develop and construct.

The following procurement methods are in declining order of control for the architects. The *traditional procurement method* is used when the client first approaches an architect, who then produces outline drawings which are costed by a quantity surveyor. The quantity surveyor may well have been appointed on the advice of the architect, though the surveyor's role is to represent the financial interest of the client. Only after the initial design phases, including costing, is a main contractor appointed. The builder has up to this point little input and no control. A tender procedure is used to select a builder, which is often the firm offering the lowest tender price. Once a contractor has been appointed, a certain amount of control shifts to the builder, but only after key decisions have been taken. The influence of builders over the design and methods of construction to be adopted in the traditional procurement method is minimal.

Many contractors argue that they are therefore forced to erect buildings using methods they themselves would not have chosen. The issue here is one of buildability or using cost effective materials, components and methods of construction through co-operation between designers and builders. As a result, they argue, buildings are often built using inefficient methods, thus causing delays and increased costs.

Quantity surveyors have also ventured into the debate over the costs and delays of construction, seeing the construction process as a management process concerned with the control of costs at both the design and construction phases. They have argued that *construction management procurement methods* are beneficial to the client because quantity surveyors are best placed to communicate the client's financial requirements to the building team and ensure compliance by both the designers and the builders. With construction management methods, clients approach professional quantity surveying practices with a brief. Later, an architect will be appointed to carry out the client's brief on the advice of the quantity surveyor. At the same time, contractors may be consulted to ensure that construction methods are appropriate. The quantity surveyor remains in control. The early involvement of a contracting firm and its familiarity with the project can give the builder an advantage over competitors at the tender stage, especially where construction methods known to the builder have been selected.

Management fee contracting further professionalises the contractor. With management fee contracting an architect or quantity surveyor, or indeed a builder, may be approached by the client. In this form of procurement method all three work together on the design and plan the construction phase. However, the management fee contractor undertakes only the professional management role on site, all work being carried out by specialist firms. This has the advantage of bringing a builder's experience in at the very beginning and speeding up the initial design stages. Indeed, the construction phase can often be started even before drawings are complete. In this method, the contractor has taken on a greater amount of influence at the early stages and hence a greater degree of control over the whole design and building process. At the same time, by not undertaking to carry out any of the actual work on site, the management fee contractor has minimised its own risk by off-loading risk on to the various specialist contractors and suppliers. The only risks for the management fee contractor involve the pure management of the site. Provided documentation is correct, the management fee contractor can locate fault elsewhere. If there are problems, it is relatively easy for the different professional practices and trades to find another firm on site responsible. Indeed, if clients wish to sue, they often need to sue several firms involved in a project, and the outcome is often far from certain.

Partly because of the difficulty, complexity and expense of pursuing litigation in the courts, a procurement system known as *design and build* has emerged as a popular method for clients. With design and build, the client approaches builders who then appoint architects to design the brief. The architects are then obliged to carry out the wishes and requirements of the builder. In design and build contracts it is the builder who assumes full responsibility for the construction process. With design and build contracts it is the contractor who is in control of the process, but who also bears the responsibility. In fact, although some larger builders employ their own architects directly, many design and build contracts are carried out by builders and separate professional architectural practices. Nevertheless, the builder remains fully in control and responsible for a given scheme.

Design and build procurement represents an added burden of responsibility for builders, whose profit margins have been squeezed since the end of the 1980s. Difficulty in finding work has forced many to lower the terms of appointment they are willing to accept. The cost of tendering for design and build work is greater than for the other procurement methods and the risks of successful litigation against them imply that contractor's design and build gross profit margins will need to increase making design and build more expensive than alternative procurement methods. Moreover, contractors will wish to avoid the greater risks implicit in design and build contracts, when their workloads increase and they are in a stronger bargaining position.

Construction markets

The construction industry consists of a large number of firms organised in distinct markets. Markets are defined in terms of buyers and sellers, the products and services on offer and the prices charged. In the previous chapter, demand was defined as the amount buyers were willing and able to purchase, while supply was the quantity of goods or services firms were willing and able to offer for sale. In neoclassical economics, the price mechanism describes the interaction of demand and supply.

In Figure 5.3, the supply and demand curves of the previous chapter are combined in one diagram. The price level at which supply and demand intersect is known as the equilibrium price. Thus, at equilibrium price P_e both the quantity firms are willing to supply and the quantity people are willing to purchase are equal to Q_e units. At prices above the equilibrium point in Figure 5.4, say P_2, there are surpluses or excess supply. The surplus is the difference between the quantity firms supply and the quantity buyers are willing to purchase. At price P_2, the size of the surplus is the difference between Q_4 and Q_2. When there is a surplus, price tends to go down. Firms reduce their prices and output. Meanwhile, as prices become cheaper, buyers increase the quantity they purchase. Price does not drop indefinitely but, as the downward

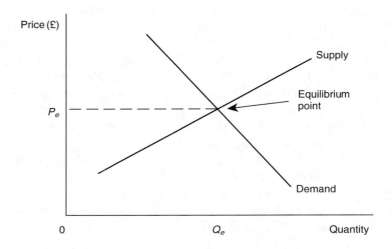

Figure 5.3   The price mechanism showing the equilibrium price

vertical arrow shows, only to the equilibrium point at P_e, where the quantities supplied and demanded are equal.

Below the equilibrium price at, say P_1, shortages or excess demand will exist. Again, price will tend towards the equilibrium level. Rising prices reduce the size of shortages in markets. The rise in price reduces the quantity of effective demand. The actual causes of shortages and surpluses can be seen in the variables causing shifts of the supply and demand curves discussed in Chapter 4.

From this simplified analysis, it can be seen that a price drop is evidence of a surplus in the market. Similarly, market prices rise in response to shortages. This focus of attention on markets is, however, a rather narrow point of view, since many people are excluded from markets when prices rise. Thus, when prices rise, shortages in the market may be eliminated but this is achieved at a cost. The cost is that the lowest-income families can no longer afford the product, and leave the market. For example, when house prices rise, it is because there is a shortage of housing. As prices rise, according to this theory, more houses are supplied and demand goes down. However, higher prices do not solve the problem for low-income homeless families. Indeed, higher house prices will tend to increase homelessness, when in reality what is needed is a greater supply of cheap accommodation. This is an example of market failure, and demonstrates the necessity for government intervention in the housing market.

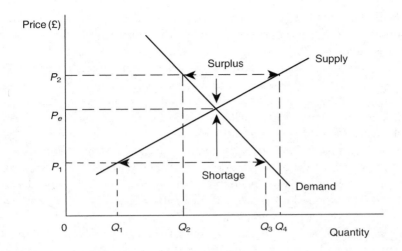

Figure 5.4   *The price mechanism showing a surplus and a shortage*

Nevertheless, as a generalisation, there is one price, called the *equilibrium price*, at which the quantity offered for sale is equal to the number or quantity people are willing and able to buy. This price is called the equilibrium price because the forces of supply and demand are balanced or in equilibrium, when there is no pressure on the price to rise or fall. This approach to the price mechanism sees outside forces preventing an equilibrium from occurring. However, it does not account for changes in these outside influences on markets. Nor is it possible to apply this theoretical approach to all construction and property markets. For example, it is not possible to know in advance, or even measure, the quantity of construction services required on a particular building project. Different firms will quote different prices to do the job.

In other markets for products and services there is the added complication of product differentiation. Every product or service provided by different firms is made deliberately different or even unique, compared to the products or service provided by other firms offering similar products. This can be seen in construction, for example, where different firms offer similar products, such as plumbing or electrical components, each with distinct features. Moreover, each office block or house is unique in terms of its location and therefore, in many cases, it will not be in competition with other offices or houses. As a property can be a one-off, it is not always possible to give it a definitive price. The price finally reached depends on the circumstances of the buyer and the seller and their negotiating skills, rather than on any market price mechanism based on supply and demand theory.

Nevertheless, markets exist in construction. Although there are many thousands of construction firms throughout the country, the construction industry is fragmented. In any one region there are so many submarkets that in any one submarket there are relatively few firms that dominate the supply of the product, material or service. Markets with only a few sellers are called *oligopolistic markets*, and the main characteristic of oligopolies is that firms avoid price competition. The objectives of firms in oligopolistic competition are to maintain or increase share of the market and maintain supernormal or above-average profits. They may therefore compete for market share by making their product or service distinct from its competitors by, for example, offering a superior after sales service than their competitors, or faster delivery, or longer credit periods. In construction, for example, a contractor may offer a maintenance contract after completion.

Alternatively, because there are only a few firms, it is possible for them to meet and get together informally to form cartels and raise prices and profits above competitive levels. Although price rings are illegal, deterrents have been minimal and supervision has been ineffective. Ready mix price rings are well known, but other cartel arrangements have come to light. For example, contractors have been known to meet to allocate winning tenders to each other. The firms tender, arranging for one of their number to tender at a lower price than the rest. The lowest tender wins, but the price charged remains above the price that would have been charged had the firms been in genuine price competition with each other. Such practices are known as collusive tendering. One method used to counter collusive practices is to insert a provision into tender documents for the examination of contractors' cost prices. In fact, the profitability of firms involved in cartels is generally low, as many of the firms are vulnerable and only survive because of the cartel practices. Others lack incentives to be efficient. Of course, some members of cartels are highly profitable.

Oligopolies and cartels exist not only in construction but across the whole spectrum of industrial society. Indeed, virtually all markets for manufactured or processed products, from petrol to computers, are dominated by a few major players, and the same is true of services, from banking to grocery shopping. The top few large firms in each of these markets account for over 80 per cent of sales.

The construction industry is a network of interrelated markets for specialist builders, equipment hire, labour, materials and components. Many different submarkets interact to produce civil engineering projects and buildings on sites. There are markets for trades, in which firms offer their services directly to the public or to other firms in the construction industry. Working for another firm in the construction industry is called subcontracting.

Separate markets have evolved in each region of the country. Unlike other industries, which tend to establish themselves in the most appropriate low-cost locations for their operations based on proximity to raw materials or other supplies or proximity to their markets, the construction industry is always market-located. Contractors must therefore compete in local labour markets for available workers, or they must transport workers into the area if the wages of imported labour plus transport are less than local wages.

Labour markets

The labour market in construction is divided in several ways: first according to trade and skills, as shown in Figure 5.2 on page 00; second according to staff and operatives; and third according to employee status or self-employment.

Labour markets are used to allocate people to jobs. In the construction industry, labour markets are fragmented according to region and the type of work or skills required. It is largely an unstructured and relatively unorganised market which relies heavily on a relatively untrained workforce. Nevertheless, the distribution of all employees according to their trade is shown in Figure 5.2. Most noticeable is the proportion of construction employees working for painters, plumbers, carpenters, electricians and general builders. However, their numbers do not reflect the number of firms, which appear to be greater than their proportion of the workforce and their contribution to output. This is because many of these individuals work in very small firms, often one-person businesses. These trades are not capital-intensive and therefore it is relatively easy and inexpensive for experienced workers to set up in business for themselves. Work in construction is therefore carried out by people who are either self-employed or employees in construction firms.

In general, there are two categories of employee in construction: staff and operatives. Administrative, professional, technical and clerical staff (APTC) tend to be located in offices, while operatives tend to be located on site. Operatives may be employed or self-employed. Figure 5.5 shows that the number of employees in employment in construction declined during the 1980s. Moreover, the decline was most marked among operatives employed by contractors. Although the number of operatives employed by contractors declined, the number of contractors APTC staff remained more or less the same throughout the period. It appears that the same number of adminis-trative staff were overseeing fewer and fewer operatives on site. Meanwhile, the number of APTC staff employed in public authorities (PA) declined, as did the number of operatives. As a result, the ratio of APTC staff to operatives in public authorities remained constant throughout the period. There appeared to be more administrators per site operative in the private sector compared to the public sector.

However, the real changes in the labour market can be seen more clearly in Figure 5.6, which shows that the decline in operatives directly employed by contractors was accompanied by a rise in the number of individuals classed as self-employed: workers were leaving employment to take up self-employment. The ratio of APTC staff to both employed and self-employed operatives remained the same. In other words, in the private sector, site labour became increasingly casualised during the 1980s.

By employing labour directly the employer maintains control over the production process. For example, the contractor knows the abilities of the workers, where they are working and what they are doing. It also enables

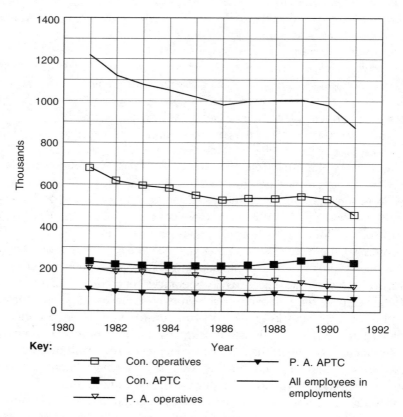

Source: *Housing and Construction Statistics,* 1992, table 2.1.

Figure 5.5   Employees in employment, contractors and public authorities, 1981–91

contractors to direct labour to specific jobs, without referring to a foreman employed by a subcontractor. Direct labour increases the day-to-day (but not the long-term) flexibility of a contractor. Directly employed labour can be more easily motivated, because employees have an extra stake in the success of their firm, namely job security and continuity of work, as well as promotion. Employing labour directly enables the contractor to get to know the workers, which improves working methods, communication and quality. If a firm gives an employee training, it follows that the benefits of the training are used in-house if the trained labourer is then employed directly.

By employing casual labour, employers avoid National Insurance costs, and other expenses such as redundancy payments. These costs are then taken on by the state, so that other industries indirectly subsidise the construction sector in its efforts to reduce the cost of building. This situation is forced on to individual contractors because of the situation in which they find them-selves. Thus, if an individual contracting firm carried higher costs than its competitors, it would need to mark up its costs at a higher rate, losing work to other firms. In fact, while it appears that contractors are being subsidised, it is in fact clients who benefit through lower construction costs. The profit margins of contractors are squeezed by competition.

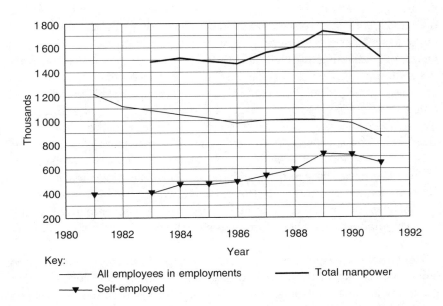

Source: *Housing and Construction Statistics,* 1992, table 2.1.

Figure 5.6 Employed and self-employed in construction

Self-employment in the construction industry can be broken down into real self-employment and labour-only subcontracting (LOSC). Real self-employment is a combination of factors. Someone can be said to be self-employed when they own the machinery and buy in the materials they use, although occasionally materials may be supplied by someone else. The price they charge must therefore, in general, cover the cost of plant, materials and labour. Jobbing plumbers, electricians and painters who, for example, serve the general public, are self-employed working proprietors.

Labour-only subcontractors are not truly self-employed, because they do not supply their own materials, nor do they use their own plant and machinery. They are classed as self-employed for tax reasons. They find employment on building sites, work with the plant and materials provided for them by the main or subcontractor, and are taken on to carry out specific jobs. They are paid on a piece rate basis called a lump sum. When their tasks are completed, their employers have no contractual obligation to give them more work. They are construction labour without a contract of employment.

The current registration system began in 1978, when tax certificates called 714s were introduced by the Inland Revenue. The self-employed status of a holder of a 714 certificate means that the labourer is responsible for paying tax and National Insurance. This system replaced an informal system of working called 'the Lump', in which contractors paid operatives directly and thus operatives were able to evade taxation. Construction labour tax evasion had become a major problem with the growth of casualisation in the construction labour market during the 1960s and 1970s.

Labour-only subcontracting can be used by employers to motivate workers and raise labour productivity in the short run. Casual employment also gives employers flexibility, as labour can be hired and fired according to requirements. The LOSC system enables employers to reduce the total cost of employing labour. They do not need to pay labour when it is not working. The contractor does not have to pay for sick leave, holiday pay, waiting time payments and wet time remuneration or pensions. Many industrial injuries in construction are cumulative, but if an injured worker has been taken on by several different contractors as a self-employed labourer over many years, it is virtually impossible to obtain compensation. Training is the responsibility of the labour-only subcontractor.

Nevertheless, casual work undoubtedly appeals to many construction labourers. As part of the culture of the industry, there has always been a tradition of itinerant building workers, and self-employment has often suited migrant labour as the only way of obtaining work. Although many operatives enjoy the freedom and independence of labour-only subcontracting, they are in economic terms casually employed labour working on poor conditions of employment. As self-employed members of the workforce, their bargaining position is determined by day-to-day changes in the labour market as well as progress on site.

Through the casualisation of labour in the construction sector, firms, especially main contractors, are separated from the site labour force. Not only is there a turnover of firms during a project, but also within firms workers leave to take up work elsewhere at very short notice or even no notice at all. Control of the labour force by main contractors is therefore minimal. Main contractors employ subcontractors, who in turn employ a high proportion of labour-only subcontractors. As main contractors do not appoint the LOSC directly, it can therefore be argued that management has abandoned control over the labour process. However, others have argued that LOSC increases control over labour, since wages are not paid if the work is substandard.

The construction labour market can be seen as an example of the general theory of the dual labour market. In this theory, the labour market is divided into primary and secondary sectors. In the primary sector, individuals are employed on a permanent basis. They have contracts of employment giving them agreed rates of pay and standard conditions of employment, including holiday pay, sickness benefits and pension rights, and they enjoy a degree of job security with career and promotion prospects. In construction, workers employed in the primary segment of the labour market have continuity of employment as long as the employer remains in business and wants to employ them. They often also have trade union protection. Their remuneration is based on time worked rather than output produced, and is usually paid weekly or monthly. In the secondary labour market, people are employed on a casual basis, with little or no job protection. They are usually paid on a piece-rate basis, often on low rates of pay. They are given little or no training and have few prospects of promotion and little job security.

However, the 1995 and 1996 Finance Acts changed the rules affecting 714 certificates. As a result, it may well be that in future, contractors will be obliged to employ a greater proportion of their workforce directly rather than on a casual basis.

Unemployment

There are various types of unemployment, distinguished from each other by their causes. The two major categories are *frictional* and *structural* unemployment. Just as friction occurs when two surfaces rub together, frictional unemployment arises when organisational changes take place between firms: for example, when two firms merge and as a consequence some employees are made redundant. The merging of companies is part of a continuous process of industrial reorganisation, needed to meet changes in the marketplace. Frictional unemployment occurs when the number of job vacancies is greater than the number of people out of work. This type of unemployment may occur even during periods of high employment. Hence, people are often able to find work in the same industry as their previous employment. Unemployment is usually short-term requiring no basic retraining of the individuals concerned.

More significant in recent years has been structural unemployment. Structural unemployment occurs when whole industries decline rapidly. The number of vacancies in these industries is then less than the number of people looking for work in them. Examples of structural unemployment have occurred in the clothing and footwear, shipbuilding, machine tool, steel and coal industries, which have all witnessed a decline in the numbers employed. The numbers employed in construction have also declined significantly.

In the period from 1976 to 1985, employment in manufacturing industry dropped by 25 per cent. It is unlikely that employment in the older manufacturing industries will ever return to the levels reached in the mid-1970s. However, opportunities in other manufacturing and service industries emerge continually as the introduction of high technology opens up new opportunities, new sources of raw materials are discovered, and business confidence returns. The manufacture of electrical goods and processed foods, North Sea oil, health, education and financial services expanded during the 1980s, increasing the number of people employed in those sectors of the economy. Since 1989, recession has increased unemployment and slowed down growth throughout most of the economy.

Because many manufacturing industries have tended to locate in specific regions, declining industries have usually resulted in high levels of unemployment in particular areas of the country. Not only do people made redundant need to retrain, but finding new work may also involve them in leaving their homes and familiar surroundings. However, the ability of labour to move to new areas of employment is constrained by the high cost and availability of housing in areas of economic expansion, where work may be plentiful. Shifts in employment patterns thus have implications for the construction sector.

Wages are not controlled by individual firms, but are the result of complex interactions between firms and those seeking work. Structural unemployment occurs if the number of unemployed people is greater than the number of job vacancies in a given industry. In a locality where the ratio of the number of unemployed to the number of vacancies (the U/V ratio) is high, wages will be low, because employers and workers will all be aware of the number of people chasing each job. Conversely, where the U/V ratio is low, wages will tend to be relatively high compared to other regions. For this reason, the wages offered for the same type of work can vary from region to region.

Plant hire market

In construction, as we saw earlier, labour is employed increasingly on a casual basis. Similarly, plant and equipment is hired as and when required rather than bought and owned by contractors. Plant hire is quite distinct from plant leasing and hire purchase, both of which involve payment in full by instalments over a period of time. Contractors often hire out spare equipment to other contractors. However, a plant-hire market has evolved, with plant-hire firms purchasing equipment and supplying it on short-term hire to contractors. Excluding inter-company hiring by one contractor from another, between 70 and 80 per cent of equipment on site is hired from plant hire companies. The proportion of plant hired in the UK is far greater than on the mainland of Europe. In 1991, the size of the plant-hire market was £181.3m, which is just under 2 per cent of the value of all work done.

In 1991, the plant-hire market was served by almost 4600 firms, employing 20 300 people. Over half of these firms employed only one person and only twenty firms employed 115 people or more. Many small plant-hire firms do not survive beyond a few years trading. As a result, although the total number of plant-hire firms remains relatively stable, there is a high turnover of firms in the market. The Construction Plant Hire Association has 1000 member firms, of which only thirty-two are subsidiaries of contractors. Indeed, although plant hire is important to construction, the largest plant-hire firms also hire plant, such as fork-lift trucks, ships' cranes, opencast mining equipment, and forestry equipment. Because they hire plant to other sectors of the economy, the construction industry represents less than half of their turnover, although they are included in construction industry statistics, such as those illustrated in Figure 5.2 on page 106.

Although hiring plant is an important strategy for contractors, not all plant is hired. In 1990, total investment in plant and machinery by the construction industry was £438m. Figure 5.7 shows the rise and fall in investment in plant and machinery by the construction industry between 1981 and 1990. Most noticeable is the steep rise in investment between 1986 and 1988, peaking in 1988 and followed by the peak in construction output two years later. It is also worth noting that the growth in construction output was not accompanied by a rise in investment between 1982 and 1987. The investment in plant and machinery was not only sufficient to replace worn-out equipment, it also

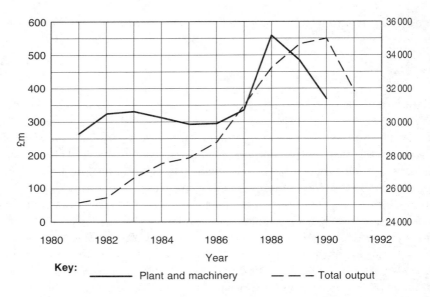

Source: *Housing and Construction Statistics,* 1992, tables 5.5 and 1.6.

Figure 5.7   *Gross domestic fixed capital formation by the construction industry and construction output, 1981–90, at constant 1985 prices*

added enough capacity to permit an increase in output. This is consistent with the theory of the accelerator principle discussed in Chapter 3.

The main advantage of hiring rather than buying equipment is the flexibility this gives to contractors, who do not need to be burdened with plant and machinery they no longer require after a project is completed. Moreover, they are not forced to spend large sums of money on equipment which would adversely affect their cash flows. The flexibility of plant hire means that firms can tender for work, knowing that they can hire the plant if they win the tender. If their tender bid is unsuccessful, then no expenditure has been incurred. Hired plant enables firms to avoid high storage, maintenance and depreciation costs, especially when the equipment is not being used. Finally, specialised plant, which might only be needed occasionally by any one contractor, can be moved from contractor to contractor and thus be put to greater use.

The disadvantage of plant hiring and casual labour is that contractors tend to become increasingly remote from the construction process itself. If they do not own the machinery they use and they only employ people on a short-term basis, it becomes extremely difficult for contractors to develop new methods and building techniques. As a result, productivity improvements may be slower in construction than in those other industries, where plant is owned or leased and people are employed and trained by their firms.

Materials

In contrast to the ephemeral nature of the labour market in which a day's lost work is lost for ever, one of the most important features of the market for building materials is that they are durable and can be stockpiled by builders or builders' merchants. Materials used in the construction process range from raw materials to prefabricated components. Raw materials include cement and aggregates, while prefabricated components include curtain walling and bathroom fittings. However, most materials entering a building site have been processed in one way or another. Steel beams, bricks and timber frames are also therefore prefabricated. Increasingly, prefabricated components are being produced for specific sites or projects. In contrast, raw materials are not site-specific, but can be used on many alternative projects.

As a result of the distinction between site-specific components and general application components, two distinct markets for materials can be seen. Builders' merchants provide the basis of a marketplace for general building materials, especially for small to medium-sized builders. The Builders' Merchants Federation (BMF) represents around 450 firms with a total of over 2250 branches. While the number of firms has reduced by more than half since 1973, the number of branches has remained relatively constant. They hold stocks of more than 40 000 lines, which are divided into heavy side and light side products. Heavy side goods are products used externally, such as roofing materials, while light side products are mainly for use inside buildings, such as heating components.

Often it is more efficient, and therefore cheaper, to buy materials from a merchant because of the saving in transport costs when a variety of materials can be delivered together rather than separately in small lots from different suppliers. Even on large sites, the administration of deliveries from one builders merchant can simplify management on site. However, because firms also buy directly from manufacturers, only around 25 per cent of the total market for building materials is purchased through builders merchants. According to the BMF, the total market for building materials has been estimated at over £22bn per annum, of which builders' merchants account for approximately £6bn. An important aspect of the market for materials that builders merchants provide is credit. Small and medium-sized builders often rely on credit until they are paid by their clients. At any one time, around £1bn is owed to builders merchants by builders.

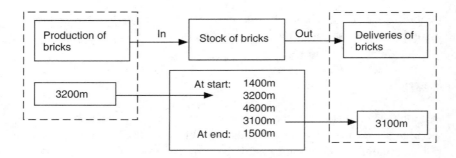

Source: Housing and Construction Statistics, 1992, table 4.2.

Figure 5.8   Relationship between stock, production and deliveries of all bricks, 1991

Because materials can be stockpiled, reserves of materials depend on production and deliveries, as shown in Figure 5.8. Production and deliveries are flows of goods entering or leaving stock in a given period. Thus, for example, in 1991, the stock of all bricks at the beginning of the year was 1400m. During the year, total brick production was 3200m, which would have raised stock levels to 4600m bricks. Deliveries from stock were 3100m, leaving 1500m bricks at the year end.

If stocks are growing, suppliers can reduce production, increase sales (by lowering their prices), or continue in production allowing stocks to expand. If stocks are falling, suppliers may need to reduce the quantity these firms want to have delivered. To achieve this reduction suppliers can either raise prices or ration supply by limiting the quantity each firm may purchase.

With general application materials and components, production usually takes place prior to sale and always before payment. With site-specific pre-fabricated components, the market operates as if the manufacturer or supplier (usually a firm not classed as being in the construction industry) were a subcontractor. Indeed, manufacturers may also provide the teams which assemble their components on site. Such firms are called *supply and fix subcontractors*. Having won a tender to supply a prefabricated component for a particular project, only then does production take place.

Land

In the market for building sites, builders and developers are buyers, and land-owners are sellers. Although there are many pieces of land available for sale at any time, each site is unique and as a result the landowner is in a strong position, if the site becomes desirable or available for development. In either case, the land owner need do little. The desirability of a site often depends on changes taking place in the surrounding area, such as improvements in the infrastructure: for example, new roads, schools or other public facilities. A site can become available for development once planning permission is given.

Land submarkets determined by location. They are submarkets because, although different sites might appear to be in competition, each site has a separate set of buyers. These buyers form competitive hierarchies of potential users. Landowners are therefore in a position to play off one buyer against another. A kind of auction can then take place as different developers compete for a site in order to put it to perhaps slightly different uses. The use that generates the highest return enables its developer to bid the highest price for the site. The value of the site is therefore determined by the use to which it will be put. There is nothing inherently valuable about a particular site. Its value depends on the ability to build a structure that will command a rent or enable an owner-occupier to generate profits. If the building is eventually owned by an owner-occupier, a rent is imputed. An imputed rent is the rent that would have been paid for the building if the building had been rented.

Thus, the value of a site depends partly on the best use to which a site can be put. Developers therefore compete for sites by finding uses for sites that generate the highest possible actual or imputed rental income. However, the price of the site also depends on the cost of construction. The higher the cost of construction, the less will be available for developers to bid for the site. This method of calculating the bid price is called the residual valuation method. This will be covered in Chapter 12. The point here is that the cost of construction has no influence on the value of a building but the lower the cost of construction, the higher the value of a site, since the savings in construction costs can be used by the developer to increase the bid price for the site. Any developer who does not pass the savings of the costs of construction on to the landowner of a site runs the risk that a competing developer, who also uses a reduced construction cost method, might do so. The significance of this for construction is that cost reductions in construction do not lead to lower building prices, but to higher land values. As a result, the price of buildings remains unchanged and there is no increase in the overall demand for construction as a result of improvements made by contracting firms.

This is not to say that land prices cannot fall. Planning blight and developments with adverse effects on a local community may lead to falls in the price of land in a particular area. However, overall, in the long run, land prices tend to rise in line with growth in the economy because of growing population pressures and the rise in productivity due to new technology and methods of production.

Speculative builders tend to stockpile land. Their stock of land is called a *land bank*. Speculative house builders tend to own land banks for two reasons. First, they need to ensure that they have a supply of land in order to have continuity of work. Because of the need to obtain planning permission, they may own three to five years' supply of land for this purpose. The second reason speculative developers hold land banks is that they buy land when it is cheap and sell when it is expensive. This is an example of pure speculation. In fact, it is the timing of the purchase and sale of land that is critical for the profits of speculative developers. For this reason, speculative developers often concentrate on land deals and employ contractors to build for them.

Social and political issues

Because of privatisation, much land and property has moved from the public to the private sector. It therefore moves into the land market, giving individuals rather than the public sector windfall profits as sites are developed. Second, public spaces such as streets are being replaced by shopping malls and other types of enclosed areas. These malls are not public spaces in that they are owned by private firms, with the right to exclude people who are deemed to be undesirable. This alters the relationship and rights of the public access to activities it might in the past have taken for granted.

In property and construction markets, prices are determined by forces of demand. Construction firms simply respond to changes in demand. For this reason there are frequent attempts to stimulate demand by putting pressure on the government to invest or create an economic climate which encourages firms to invest in civil engineering projects and building proposals. They point to unemployed construction workers alongside homelessness. At the same time, many office blocks produced by developers remain empty. The economic and political issue here concerns the use of scarce resources to produce buildings which become surplus to requirements instead of providing housing that was needed.

Turning from property issues to construction, there are a number of economic issues arising from the response of firms to low levels of demand because of recession (as shown in Figure 3.8 on page 52), a persistent feature of the construction market in the first half of the 1990s. Many firms claim that over-capacity in the industry is driving prices down towards uncompetitive levels. However, in any highly competitive market there will always be the appearance of too many firms chasing too little work: that is the nature of competition. In this way, according to the neoclassical theory of competition, the least efficient and least profitable firms are driven out of the market.

However, in construction, this is not necessarily the case. In construction the tendering process means that normally the lowest tender price wins, but often the most efficient and best practice firms do not tender below the price they think they realistically require in order to make a profit. By bidding in this way they can be undercut by other firms willing to push their suppliers' prices down, especially after a contract has been won. Best practice may mean paying higher wages to skilled workers, and providing higher standards of training, and higher levels of safety and investment in research and development, than those firms putting in the lowest tender price in order to win contracts. It is therefore not surprising that firms attempting to survive in the short term do not necessarily perform well on site. Low tender prices force contractors to take short cuts, such as employing cheap, inexperienced casual labour to carry out tasks requiring skill. As a result, things frequently go wrong and the matter ends in disappointment and litigation.

This is not to be blamed upon the firms, who are forced by competition to take desperate measures to win work, and low profit margins mean that they have to win even more work in order to generate sufficient profits. This further drives down prices and profit margins, with the consequence that firms run into cash flow difficulties, sometimes ending in liquidation during a project. The problem for firms is that they operate within a system of building procurement and in competitive markets which no one firm can alter. The arrangements that exist in construction have risen for historical reasons, discussed in Chapter 2.

To a great extent, these arrangements are based on three factors. The first is the traditional separation of design from production. Design is carried out in professional architecture and engineering practices, whereas production is undertaken by contractors and subcontractors on site. Second, there is a wide range of procurement systems. These procurement systems meet the need for flexibility caused by the complexity, unpredictability and discontinuity of work. Third, the operation of the tendering system is designed to obtain the lowest-price contractor for the client.

Investment in expensive plant and machinery can often only be justified if it can be used more or less on a continuous basis. This is unlikely to be guaranteed, except for general equipment such as dumper trucks and cement mixers, as main contractors take on a large variety of work, using different methods of construction. It is invariably easier and cheaper for firms to hire more specialist plant and equipment than to buy it. As Ball (1988) has pointed out, it is essential for firms to remain as flexible as possible in order to be able to respond to the different kinds of work they may be asked to carry out. The variety of work using different types of plant and labour reduced the incentive to invest for many firms in the construction industry. Plant hire and casual labour became the norm in the 1980s. It was also cheaper for firms to lay off workers immediately after a project was complete. The tendering process had worked to reduce prices and mark-ups to a minimum, so that any idle labour that required to be paid following the completion of a project would quickly use up the firm's profit margin. In other words, the working conditions of labour are the ultimate consequence of the competitive nature of the tendering and contracting system.

Each firm must participate in the contracting system in order to obtain work and survive. Those firms willing to offer the lowest tender prices have tended to win the work and, in order to remain flexible and competitive, they have hired and subcontracted. Once they have been appointed to carry out work, they are in a position to maintain their profit margins even in a recession. After winning a contract, the main contractor can return to the subcontractors, who have helped in the preparation of the tender, and require them to reduce their prices. In turn, subcontractors reduce the wages they offer to labour in order to maintain their profits. As materials and other costs cannot be altered by management as easily as wages, labour is targeted as the most accessible cost-cutting element, especially in periods of high unemployment.

Main contractors act as management agents, controlling payments. As main contractors, they are in a strong bargaining position, since they tend to be larger than their subcontractors and can hold the offer of potential work in the future as an implied incentive to encourage subcontractors to co-operate. With the reduction of their own workforces on site, replaced by specialist subcontracting firms and labour-only subcontractors, the trend has been for main contractors to undertake less and less of the actual building process. This trend has been reinforced by the increased use of prefabricated components. Indeed, with management fees contracting, the main contractor has shifted to the professional role of management, directing specialist contractors to carry out almost all of the work.

Indeed, since M. Bowley wrote in the 1960s (Bowley, 1966), the changes in procurement systems have reflected changes in the roles of architect, quantity surveyor, contractor and subcontractor. With design and build contracting, contractors replaced architects in the lead role. This procurement method allows the contractor to instruct the architect, which is the reverse of the traditional procurement system. With design and build, the architect can hardly claim to represent the client as it is the builder's choice which architect to appoint.

There are many other tensions in the construction industry, apart from those between the different professions. The other areas of tension are between firms and employees, between main contractors and subcontractors, and between contractors and employers. To understand the nature of the causes of these tensions it is useful to look at the economic nature of the production process itself.

Surplus

Let us build an economic model of the production process. The value of output is greater than the value of material inputs. The difference is called value added, because the value added to material inputs plus the cost of these materials is the value of the production process. The value of output must be shared between the workers who produce the output, the materials used up in the production process and the owners of the plant and equipment used. If the amount paid to workers and the price paid for materials were equal to the value of output, then nothing would remain. There would be no surplus. The surplus is used by the owners of the firm to finance investment in machinery as well as the dividends paid to shareholders. If there were no surplus there would be no money to pay rent to landlords or interest to banks. In other words, surplus is the source of funding for rent, interest and profits.

In order to ensure that there is a surplus, wages rates must be held down. Consequently, there is a conflict of interest between the owners of a firm and its workforce. The higher the wages, the lower the surplus. However, the conflict does not end there. Rent is paid to landlords, interest is paid to lenders, and profits are paid to the owners of the firm, only *after* rent and interest have been deducted from the surplus. The more paid in rent and interest, the less is left for profits. Hence, there is a conflict of interest between landed capitalists, banking capitalists and factory-owning capitalists. Similarly, the firms involved in the construction process must compete with each other for a share in the surplus. The more that is paid to one firm, the less to another. For this reason, firms and professional practices in the construction process vie with each other over control of the process. They are in constant conflict over payments, the share of work on a project, and the allocation of risk.

The construction industry is seen by many as being adversarial and litigious. That is, conflicts between participants in the construction process frequently end up in court. One reason is that contracts cannot possibly predict all eventualities, in spite of the wide variety of contract forms. When something occurs outside the scope of a contract, the parties may resort to the courts to resolve the dispute because neither side is willing to concede. Making concessions is difficult because profit margins are often so low that a concession could wipe out any profit. The reason for the persistence of low profit margins is the highly competitive nature of the tendering system and the tendency for the lowest tender to be selected.

This high degree of competition places additional pressure on firms and leads to a blame culture in the construction industry. One of the main areas of conflict in the construction industry is between main contractors and subcontractors. The division of the production process between contractors and subcontractors leads to fragmented management, which in turn leads to poor co-ordination between design and other consultancies. Moreover, because of the high cost of insurance, firms are discouraged from accepting liability, especially where an insurance claim may result.

However, it is on the subject of payment to subcontractors where many disputes arise. The system of payment is based on 'pay when paid' clauses. In other words, main contractors undertake to pay subcontractors only once they themselves have received payment from clients. This avoids cash flow difficulties for the main contractor while placing an additional burden and risk on subcontractors. If, for any reason, a payment is held up, payments to subcontractors can be delayed, even when the subcontractors have performed satisfactorily. Often there is little that subcontractors can do, since in order to sue a client for late payment a court would require to know not only how much was owed but also the precise date on which the payment became due. Pay when paid contract clauses mean there is no exact date for payment.

It was to look at these conflicts in the construction industry that the Latham Report was commissioned, some thirty years after the Banwell Report had dealt with similar issues. Published in 1994, the Latham Report, called *Constructing the Team*, examined procurement methods used in the construction industry. The issues raised in the Latham Report included the problem of briefing, the problems of working relationships between main contractors and subcontractors, and the lack of innovation and hence the high price of construction in the UK. Latham was also concerned with guaranteeing quality and proposed new legislation to consider unfair contracts, latent defects and professional liability.

In spite of these difficulties, the Latham Report set out to create an atmosphere of mutual trust and co-operation by encouraging the use and development of new contractual arrangements, based on the New Engineering Contract as well as on partnering. However, the underlying confrontational atmosphere in the construction industry has more to do with underlying economic causes than contractual arrangements.

Who benefits from the construction process? The simple answer is that the construction workers gain employment, while the contractors and developers gain profits, and the building users receive a completed structure. However, it is important to consider the sharing of surplus value between contractors, developers and landowners. Developers must compete with other developers in order to obtain the right to build on a site. There is, in effect, an auction. The developer with the highest offer generally wins the bidding. In order to win, the developer must be able to obtain planning permission for a proposal, but proposals are limited by planning constraints. One developer's proposal will tend to be similar to the next. One way a developer can increase the offer for a site is by finding a contractor willing or able to construct the building at a lower cost compared to competing developers construction cost estimates. For this reason there is always downward pressure on construction tender prices.

As noted in Chapter 2, if a contractor can build for a reduced price, the reduction in price can be used by the developer to increase the offer price for the site. Reductions in construction costs therefore contribute to higher land prices, while the final building, including the site, remains at the same price. The contractors charge less for their work, but because the cost of buildings remain the same, there is no increase in overall building work as a result of the reduction in construction costs. The reduction in construction costs does not benefit developers either. They only use the reduction to increase their bids for sites. The ultimate beneficiaries of improvements in construction technology and techniques leading to increased efficiency in building production and lower construction costs are landowners, who capture a larger share of the surplus produced by the building workers. This is not to say that those who work in property, development and finance do not make a contribution to the process. Without their effort, construction work would not be initiated and could not take place. Property companies which deal in land serve to allocate sites for different purposes through the market mechanism. They may have a vested interest in maintaining the property system in the form it has evolved, but they did not create it.

6 Construction Labour Economics

Introduction

Labour is central to any production process, and construction is no exception. Without labour, production would not take place. Machinery would remain idle. Stocks of materials would not be made into products. Firms would not be managed. To understand the role of labour in the capitalist production system we need to see labour in terms of its relations with the other participants in the process – the owners of capital and land.

Labour is provided by people, who are paid a wage or salary. This wage must be less than the value of the goods or services produced, otherwise there would be no money to pay for the plant, equipment or materials used in production. After all costs have been paid, the remainder is profit. The owners of the firm, usually the shareholders, receive a share of part of the profits distributed to them in the form of dividends. Those profits not distributed to shareholders are retained by the firm for future investment.

In a simplified model of property development, a landowner owns a site, a developer purchases the site and a contractor builds. The contractor has to bid for the work in competition with other contractors. The lowest bid wins and the winning contractor then buys materials, hires plant and employs workers. Workers compete for jobs but there is a minimum wage below which they are not prepared to work. This wage is known as the *subsistence wage*, and is the minimum necessary for survival, although this has come to mean feeding, clothing and housing a family at a minimum socially acceptable level.

When production takes place, materials, including energy, are used up. For production to continue, materials need to be replaced. The value of output must cover wages and the cost of replacing materials used. If there is any money left, after paying for labour and materials, then the firm is said to have a surplus. The purpose of production from the point of view of the owners of the firm is to produce a surplus. Unfortunately, the owners of the firm must share this surplus with others. If they have borrowed money to finance production, then interest must be paid. If they use land belonging to a landowner, then rent must be paid. Only after interest and rent have been paid are the owners entitled to the residual called profits.

It therefore follows that the higher the wages, the smaller the surplus, and the smaller the surplus, the less will be left after interest and rent. Indeed, the lenders and landowners will compete with the owners of the firm for a share of the surplus. The owners share of the surplus is called profit, which is then retained for reinvestment in the firm, or distributed to shareholders. The whole system of production involves conflicts of interest between workers and capitalists, and between owners of firms, money capital and land.

The role of labour is to produce surplus value for capitalist firms. The ability of the firm to survive, to reinvest in its future, to produce sufficient profits for distribution to its shareholders and to pay interest and rent depends on the surplus value produced by labour. If there is an increase in the surplus value produced, then profits, interest or rents rise. As interest rates are usually determined by market forces, an increase in surpluses will tend to be shared between landlords and the owners of firms. Indeed, firms compete for premises and those firms which produce the greatest surplus are in the strongest position to pay the highest rents. Thus, the higher the surpluses produced by labour, the greater the rent that can be charged. The benefits of the production process therefore accrue to landowners over time. This model of production can be observed repeatedly in different sectors of the economy, and those locations in a city which enable firms to produce the highest surpluses command the highest rents.

Industrial relations in construction

Clearly, there are conflicts of interest between labour and employers. Labour tries to maximise its wages and employers try to drive wages down in order to minimise costs and remain competitive and profitable. There is also, however, a need for labour and employers to co-operate. As methods of production become increasingly mechanised and technically sophisticated, the construction process becomes more capital-intensive. The more complex the production process, the greater the need to be co-operative, as firms require reliable workers to operate the plant and equipment profitably. Workers can benefit from the success of their firms, with increased job security or at least the continuity of work, though this does not always happen. With experience of working in construction, and the casual and temporary nature of employment, many workers feel little unity of interest with management or the owners of firms.

The relations between management and site labour is complex. In construction there are structured negotiations between employers' representatives and the construction trade unions. The employers' representatives include the Building Employers' Confederation. The employers' representatives and trade unions meet in a number of joint bodies, the most notable being the National Joint Council of the Building Industry (NJCBI), and the Construction Employers' Federation. The civil engineering industry conducts its negotiations separately through the Civil Engineering Construction Conciliation Board (CECCB).

The National Joint Council of the Building Industry, which sets various wage rates for the industry, provides the forum for annual negotiations concerned with the rules governing industrial relations in the construction industry. The outcome of these discussions is the National Working Rules Agreement covering hours of work, working conditions and health and safety issues. Although these negotiations are held annually, the rules, wages and conditions are rarely adhered to on site. Nevertheless, the agreements reached provide employers with a point of reference. For example, firms can use agreed rates of pay to cost work, even when actual pay is at variance with the estimate. Similarly, the National Working Rules Agreement can be used as a framework for subcontract agreements.

Alongside the *formal* arrangements described above, involving representatives of employers and labour, there is a parallel *informal* system of wage bargaining taking place directly between site managers and workers on site. Rather than accepting formally agreed rates of pay and conditions, the relative bargaining positions of site managers and workers on any site depend on the organisation of the site, local labour market conditions and the shop stewards (if any are present on site), who are the locally-elected trade union organisers.

There is a third system of industrial relations, which has been called the '*unformal*' system (Druker and White, 1995). The unformal system involves self-employed labour-only subcontractors, who obtain work on site on the basis of conditions outside any agreements reached at national, or even local, level. These LOSC conditions of employment were discussed in the previous chapter and now apply to over half of the operatives working on most sites.

Wage-forms

The actual amount paid in wages is not the only concern in the bargaining process: the method of payment is also important. The terms of employment, including the method of calculating pay, give various degrees of control to the employer, especially where performance is related to pay. The method of calculating pay is therefore a focal point in the relationship between workers and employers. There are several methods of calculating pay.

The *piece-rate method* is payment according to how much is produced. In construction, piece rates are used for paying operatives for a given amount of work. The operative is thus given an incentive to finish the work and receive payment. The sooner the work is completed, the sooner payment is made. Piece work is the single most common form of payment in construction.

Labour-only subcontracting is an example of a piece-rate-based system of working applied to the construction sector. A gang of workers is given a fixed price for a quantity of work. The contractor supplies the heavy machinery but the gang supplies its own hand tools. For example, a gang of bricklayers may work for a lump sum. Similar systems of employment once operated in coal-mining, where it was called the 'buttie system', and in the docks, using the 'gang system'. In factories, internal subcontracting occasionally involves a similar method of subcontracting piece work to a small group of people. Professionals, such as barristers and consultants, work under similar conditions, in this case called 'freelancing'. These are all examples of pure-piece rate working.

This form of piece-rate labour has some short-term advantages. For example, workers tend to work intensively in order to earn their wages promptly, and this raises productivity in the short run, adding to a firm's competitiveness. However, people employed on piece rates have few opportunities to train themselves, as any time off the job means a loss of wages for the employee. Moreover, as training others costs time on site, there is no incentive to train others. Nor do firms have an incentive to invest in expensive plant and machinery, especially where a period of training of casual labour might be involved. In the longer term, innovation is inhibited and productivity growth is slowed down, (Janssen, 1991).

The *time rate method* rewards people for the time spent at work. With monthly pay, workers receive a salary even if no productive work has been carried out. With wages based on time rates, operatives have little incentive to speed up work, but instead are given time and the opportunity to produce a higher quality of workmanship than they would under the pressure of piece work, though it has been argued that if work is unsatisfactory using the piece work method, operatives' pay can be withheld. When workers are paid on the basis of an hourly rate, the working day can easily be extended, and therefore working conditions can be lowered. Moreover, hour rates increase employers' control over labour. For example, hourly pay with two hours' notice would increase worker insecurity, but from the employers' point of view, hourly rates aid control, costing and planning.

There are a great variety of *bonus schemes*. Often bonuses become an expected norm and defeat their purpose of acting as an incentive to complete work on time and to the expected quality. This is especially the case where the contribution of individuals cannot be measured. Measured day work is a combination of time rate and piece rate. The amount of work to be completed in a day and appropriate remuneration are negotiated. However, when day rates are used, employees do not have job security and can be dismissed on completion of a contract. Another problem with using day rates is agreeing the length of the working day needed to carry out the quantity of work.

It does not follow that systems of payment cannot be changed, just because wages have traditionally been paid in a certain way. In the first half of the twentieth century, Taylorism (time and motion studies), and Fordism (mass-production techniques), treated people as if they were simply motivated by money. However, when capital-intensive methods of production are used, more sophisticated methods of payment are often called for. Indeed, it has been argued that, as new techniques are developed, new forms of payment become necessary. These in turn change the relationship between employers and employees. This can be seen clearly in the car industry and in computer software services, where new technology has altered production methods, and in some cases, labour relations.

Training and skill shortages

The growth of LOSC in construction has implications for training. LOSC reinforces traditional trade demarcations and leaves no time for training new recruits. The issue is that the cost of education has to be set against the discounted flow of benefits arising from training. This means that firms must be able to recoup their training costs from the extra productivity of trained workers. At the same time, the workers have to benefit from extra income and employment, which they would have missed without training. Workers and employers alike must compare the costs associated with training with the alternative of no training. Insufficient training can adversely affect the organisation of the production process on site, the efficiency of the construction sector as a whole, and even economic growth.

In the past, apprenticeship was the main form of training for bricklaying, plastering and plumbing. However, formal training does not guarantee quality; for example, where workers lack motivation. Nor can it prepare workers for the variety of problems and materials they will have to deal with in their working lives. Nevertheless, the construction process is affected by the quantity and quality of training. Formal training is undertaken in the main by the CITB, colleges, DLOs, and some firms. However, most new entrants to the construction industry learn informally. Informal on-the-job training, such as observing, or being shown by, superiors or workmates, takes place when new entrants to construction join small teams of workers.

In the UK there has been a history of underspending on training compared to other European countries. Thus, although training in the construction industry is employer-based, firms themselves are reluctant to train casual workers, who would only take the knowledge gained to their next employer. Instead, firms rely on sufficient numbers of people applying for work when they are needed. The construction sector is largely composed of project teams, temporary organisations using an existing workforce on an *ad hoc* basis. As there is little training, skill shortages are a constant source of uncertainty for contractors. Shortages of skills lead to rising costs, reduction in product quality and difficulties in introducing new techniques, if the skills are not available. This affects the competitive position of firms adversely. Skills in short supply may be attracted from abroad, and migrant labour has traditionally come to the UK from Ireland.

Skill shortages are difficult to define and quantify in terms of market economics. A shortage of skilled labour is often the result of low wages being offered, leading to unfilled job vacancies. The number of people offering the skill increases when firms offer higher wages. Higher wages raise costs, and demand goes down as architects and clients look for cheaper alternatives or substitutes. In this way the market mechanism eliminates shortages.

Since 1960, the number of trainees in construction has declined by 60 per cent, partly due to the raising of the school-leaving age. The drop in trainees was not reflected in the drop in employment, which fell only by 16 per cent from the late 1960s to the late 1980s. Direct employment over the same period dropped by 42 per cent, further reducing the number of opportunities for school-leavers to take up training. In the 1970s, the proportion of trainees to the total labour force dropped still further, and this trend continued throughout the 1980s. Further skill shortages may occur in the 1990s, because of demographic factors and a fall in the number of school-leavers. At the same time, new technologies will require new skills and more training.

Since the 1960s, then, skill shortages have arisen periodically, with consequences for economic growth. Whenever expansion of the industry takes place, firms compete for all available labour, raising wages until sufficient people can be attracted away from alternative sites and employers. This leads to price instability and uncertainty for building employers, who want to plan and budget for their projects.

In 1964 the Industrial Training Act set up training boards, involving the government, employers and unions. Grant levy systems were set up in several industries, including construction. While all firms paid a levy towards training costs, only those firms which undertook to train new employees received a grant for each trainee. This system did not ensure that the training given was always appropriate, however, although some firms and organisations undoubtedly took their responsibilities seriously, it was difficult for the training boards to monitor all the training that was given. Because of the dispersed nature of the construction industry, the training given was controlled by the employers. Nevertheless, the system encouraged firms to provide training in order to receive grant income.

The Industrial Training Boards were weakened by the Employment Training Act of 1973, which attempted to replace the grant levy system, and in 1981, all the training boards except the Construction Industry Training Board (CITB), were abolished. The CITB survives, funded in part by the grant levy scheme. In 1983, the CITB provided 23 000 places but as these were dispersed among many building sites, it was difficult for the CITB to ensure that the training agreements were adhered to by employers. In 1995 there were approximately 30 000 construction trainees in total in one form of training or another. While the numbers appear to have grown, the quality of some of the training provided has given the industry cause for concern. National Vocational Qualifications (NVQs) have been introduced, and a new Construction Skills Certification Scheme (CSCS) has been set up to encourage improved training standards. However, throughout the 1980s trade unions were increasingly excluded by the government from having an involvement in training. In 1988, for example, the Employment Act abolished the Manpower Services Commission (MSC), which had set up training schemes with union participation. The Employment Act replaced the MSC with Youth Training (YT), in which the unions were not involved. Employment Training later extended YT to adult training and employment.

When youth wages are relatively high, employers argue that they cannot afford training costs. Hence, it was thought that if wages were reduced, then more training would be forthcoming. In fact, the Youth Training Scheme (YTS) was used to subsidise youth employment, in the hope that training provision would be increased. However, the result of YTS has been to dilute training and to give employers the opportunity to take on 'trainee' labour at subsidised rates of pay. From the point of view of training labour, the schemes have made only a small contribution, although undoubtedly some people did manage to gain employment through starting work with youth training schemes.

In construction labour markets it is not possible to know the level of skill of operatives offering their services, because of the informality of training and the lack of generally recognised qualifications. Often no specific training is required. Conventional methods are learned on the job. Experienced roofers, for example, point to the variety of problems they face, which no amount of formal training could provide. Lack of formal training does not necessarily mean lack of skill. The number of people with any particular skill is unknown. In any case, experienced building workers are continually moving into other industries and jobs outside construction, especially into furniture manufacturing, labouring and other unskilled work, and taxi, bus and lorry driving.

Because much of construction labour is under-trained and casually employed, ensuring that a satisfactory quality of work is achieved is extremely difficult. Many firms have attempted to get round the problem by adopting a quality assurance standard, called BS 5750, which since 1994 has been the international standard BS EN ISO 9000. This is a British Standards Institution Certificate of quality control given to firms that can demonstrate a quality control system of management. It is concerned mainly with the management of projects rather than labour, and involves setting up procedures and ensuring that a system of documentation is maintained at each stage in the production process. In fact, BS 5750 is designed to ensure that firms carry out precisely what they undertake to the standard and quality agreed at the outset. It is not in itself concerned with improving methods or in rectifying problems when they occur. It can, however, show where errors in communication may have been made, for example, when things going wrong. Over time, firms using BS 5750 may therefore learn to avoid repeating mistakes. However, it is aimed at management rather than operatives.

Productivity

Unit labour costs are the cost of labour per unit of output. They can be calculated by dividing the cost of labour by the number of units produced. As labour is usually the largest single item of cost per unit of output in manufacturing, the cost of production is sensitive to reductions in labour costs. The same is true in construction, and one of the main driving forces in construction labour markets is the need of firms to minimise their labour costs in order to remain competitive.

Unit labour costs can be reduced by increasing productivity or cutting wages. While raising wages will attract the most able workers, lowering them will have the opposite effect. If the best workers leave, a firm does not become more competitive. An alternative strategy for management is to increase productivity. Productivity is the output per person in a given period of time.

By improving the management of sites and the building process, it is possible to increase productivity. This is done by reducing the porosity of the working day; that is, by reducing the proportion of the working day that is wasted. Management can measure the porosity of the working day by using the ratio of productive time to paid time: the smaller the ratio, the greater the porosity of the working day.

In construction, productivity can be measured by using the value added on site. Value added is the difference between the price of a construction contract and the cost of materials, prefabricated components and other inputs, except labour. Through its labour, a firm adds value to these inputs. This added value can be used to calculate productivity, by dividing the value added on site by the number of people employed. Value added is equivalent to net output. Net output is gross output less inputs. Hence, productivity, or net output per head, is net output divided by labour. Productivity also takes the duration of the project into account. Output per head per week is net output per person divided by the number of weeks:

$$P = \frac{NO}{L \times T}$$

where
$\quad P$ = productivity
$\quad NO$ = net output
$\quad L$ = labour units
$\quad T$ = time periods

Although net output is declining because of the increased use of prefabrication, which in turn leads to the deskilling of site labour, one way of increasing productivity on site is investment in new technology and techniques. New technology tends to reduce the demand for labour, as new inventions and innovations are often labour saving. In general, on large projects, the trend is for fewer people on site working with more plant and machinery per person than in the past. In other sectors of the economy, most notably in consumer goods manufacturing industries, the introduction of new technology is followed, after a time lag, by lower prices. Lower prices in turn increase the demand for a product, which can raise the demand for labour. In such cases, new technology can both reduce labour costs per unit and raise the demand for workers.

In construction, however, new technology only tends to reduce the demand for labour, especially in the traditional skills. When construction unit labour costs are cut, the ensuing lower prices may not be passed on in the form of cheaper buildings, because the total final cost of a building is composed of land as well as construction costs. The cost of production (construction labour, materials and components, and plant and equipment) may have little influence on the price of a building. A reduction in construction costs has in certain situations the effect of raising land prices, as developers compete for available land with increased bids derived from the extra savings on construction. This will not apply where the land is in the ownership of the client or has already been purchased. However, in the long run, reductions in construction costs in general lead to higher land prices rather than cheaper buildings. Thus there is a disincentive to innovate in the building sector, since in the long run, construction price reductions are rarely passed on to the eventual buyers of completed projects.

Productivity relates to the output of any factor of production. Hence, labour productivity is a measure of the output of an individual in a given period of time. It may be measured in terms of money or in terms of what was actually produced. The productivity of an architect may be measured using the number of drawings completed per week. Of course, such a measure of an architect's output ignores the quality and variations in complexity between drawings and the time taken on each one. Nevertheless, over a sufficiently long period of time a pattern will tend to emerge, and a given number of drawings or other activities may be expected. Once an average output figure is found, it may be used as a standard. Output by individual architects can be compared to the expected standard. Where a difference in performance persists, the reasons will either be obvious or reasons should be sought.

Adam Smith recognised as long ago as the eighteenth century that it was possible to increase labour productivity by dividing work up into tasks. A number of workers could then share the work, each one specialising in his or her own task. This division of labour greatly increases output by saving time and taking advantage of personal abilities, skills and preferences, but it also entails the disadvantage of interdependence. One member of the team is in a position to disrupt the whole process and delay completion of the work. However, as Charles Babbage pointed out, individuals could just as easily carry out different tasks each day until all the tasks were completed. In this way, each person would remain independent of the others. What the division of labour between several workers meant was that control over the process shifted from the people carrying out the work to their employers, who were then in a position to organise labour to suit their purposes. Thus labour could be deskilled by being trained to carry out only one or two tasks, but not the whole process. Being less skilled meant they could be paid less and the employees became more dependent on their employers for work. Other employers would need to train them in other specific tasks, and the employees would not be able to work independently of employers.

The law of diminishing returns

The law of diminishing returns is an important theoretical economic tool of management. The return from successive units of labour is not constant, but declines eventually. This law of economics states that assuming all other factor inputs remain the same, as one increases the amount of one factor, then output will increase. However, eventually, the rate of increase will begin to decline. In Table 6.1, the total product rises steadily at first, as extra workers are employed. The average product is the total product divided by the number of workers. The marginal product is the increase in total product which results from employing one extra worker. Hence, when the number of workers rises from, say, four to five, the total product rises from 55 to 65 units, a marginal product of 10.

Table 6.1   The effect of employing extra workers on total, average and marginal product

Number of workers	Total product	Average product	Marginal product	Value of marginal product
1	10	10	—	—
2	25	12.5	15	150
3	40	13.3	15	150
4	55	13.75	15	150
5	65	13	10	100
6	70	11.67	5	50
7	70	10	0	0
8	65	8.13	−5	−50

This can be seen in Figure 6.1, where the total output curve, or total product (TP), rises as the amount of labour increases. However, there will come a point beyond which an increase in labour will result in a declining rate of return. This, too, can be seen in Figure 6.1, where the rate of increase in the total product curve begins to slow down even though the curve itself is still rising. Eventually, even the total product curve declines, beyond x units of output, in spite of increases in labour. The diagram shows the relationship between total output and the rate of increase in total output, called the marginal product (MP), when labour is increased. Where the marginal product curve cuts the x-axis at 6.5 workers, there is no increase in the total output. (If 6.5 workers seems a strange number of workers, a part-time worker could be given the value 0.5.) The marginal product measures the change in output caused by employing the last unit of labour. Where one more unit actually reduces total output, the MP curve is negative.

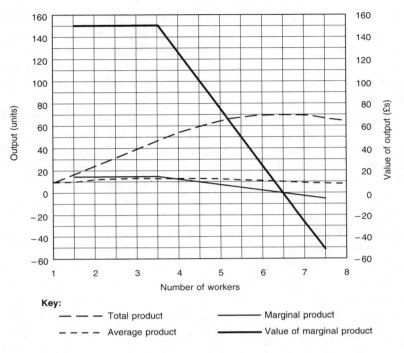

Figure 6.1   *The law of diminishing returns*

It is important to appreciate that the last labourer is part of a team, and that the total output is the result of team effort. The law of diminishing returns shows that even though the last person employed might be the hardest worker, the significant factor is the difference the last person employed makes to the total effort of a team. The question to ask when hiring people is, what difference will one more person make to the value of total output? Alternatively, the question can be asked, would the saving in wages if one less person were employed be greater than the drop in the value of output? If so, reducing staff numbers by one would increase profits. For this reason, the trends in productivity and unemployment have both risen since the 1970s.

Health and safety issues in construction

Accidents causing three or more days of absence from work must be recorded and notified. In 1993, the total of all three-day injuries in construction was 15 855, down from 29 997 in 1989. This drop in the number of injuries is partly due to the decline in workload since 1989, and partly due to an improvement in the safety record as measured by the rate of injuries per 100 000 employees. The number of fatal injuries in construction in 1993 was 87. In 1989 there were 137 fatalities in construction, and in 1990 the number reached a peak of 154. These figures do not compare favourably with other industries. Construction accident rates are around twice the rate in the manufacturing sectors.

The most common cause of fatal injury in construction, accounting for almost half of all fatal accidents is falling from a height of more than 2 metres, mainly because of speed of working and falling through roofs. Other causes of fatal injury include being trapped by something collapsing or overturning, being struck by a moving vehicle or a falling object, and contact with electricity. As a result of fatal accident enquiries, many of these accidents are later found to be a result of inadequate training or instruction, inadequate supervision, or unsafe methods of working. Faulty structures and working platforms as well as unsuitable access are also underlying causes of many fatal accidents. These are all management issues. Driver or operator errors are found to be a contributory factor in a minority of incidents, but responsibility for fatal accidents lies mainly with management.

In 1965 health and safety regulations based on a variety of Factories Acts were drawn up. In 1974 the Health and Safety at Work Act set up systems of safety representatives on site and safety committees. Inspectors from the Health and Safety Executive were empowered to visit sites. Around 600 inspectors are employed by the Health and Safety Executive. Because of the great number of building sites, inspectors can only visit a small proportion of them. Where problems are found, health and safety inspectors may advise site managers to make improvements. If the problems are more serious, inspectors can place Improvement Orders on a site. An Improvement Order is a threat to close a building site, unless changes are made to remove the danger. For serious breaches of health and safety regulations, inspectors can make Prohibition Orders, which cause work to be stopped.

New regulations were introduced in 1994. Called Construction Design and Management Regulations (CONDAM), the new health and safety measures attempt to ensure that employers and designers take responsibility for the prevention of accidents on site. The regulations require that planning supervisors are appointed to oversee health and safety matters on site. The actual role and legal position of planning supervisors is, however, not made clear, and may vary from project to project.

Control over safety aspects is exercised through a health and safety plan (HSP). The HSP takes into account the nature of a project, including the type of building work; the design, including construction methods; the site, including ground conditions, hazards and precautions required; and the management of the project, including site rules and health and safety procedures. In this way, developers are held responsible for ensuring sufficient funds are made available for health and safety on site, and that health and safety conditions are included in their contracts.

It remains to be seen to what extent the HSP is an effective device for improving the safety record of construction projects. One of the most common reactions by firms in the industry is the fear that the new health and safety measures will increase construction costs and reduce demand for construction. However, if all contractors are obliged to conform to the new regulations, then no single contractor would be disadvantaged in the tendering process. Moreover, as we have seen above, if lower construction costs can lead to higher land prices, then the additional cost of new safety measures would be absorbed by lower land prices leaving the final combined cost of the building and site unaltered. In any case, it might be argued that improved health and safety measures might actually reduce construction costs by reducing the cost of injuries and accidents on site.

In spite of legislation and the work of the Health and Safety Executive, construction work remains dangerous, partly because of the way management of sites is organised, and partly because of the nature of building and civil engineering work itself. Indeed, many workers in construction eventually suffer disabilities as a result of cumulative exposure to harmful substances, repetitive work and lifting heavy weights. As labourers and operatives may work for several contractors, it is almost impossible for them to claim compensation for industrial injuries caused by work practices carried out over many years.

In general, complaints by workers are discouraged. Competition between contractors combined with very tight profit margins mean that savings on site are made wherever possible. A further impediment to safety on site is the complexity of management structures arising out of the subcontracting systems used, making it difficult to regulate workers on site and allocate responsibility for safety. For example, insisting on the wearing of hard hats on site is frequently a problem for site managers, often because the site managers are not the direct managers of labourers working for subcontractors.

Moreover, the haphazard conditions on site make construction work inherently unpredictable and often dangerous. By their very nature, materials, equipment and construction conditions may be dangerous. Materials such as molten steel, cement, paint and asbestos contain unavoidable or potential or recognised hazards. Equipment, such as high pressure water jets (which, according to the Association of High Pressure Water Jetting Contractors, produce pressures as high as 36 250 lb/in^2 or 2548 kg/cm^2, with a jet velocity of 1536 mph or 680 metres per second), require extreme care in their operation. Scaffolding, roofing and trench digging work are all potentially dangerous working conditions.

Safety involves making situations, working conditions and work practices as safe as is reasonably practicable. This means taking precautions, training and instructing staff and operatives, and ensuring management procedures are in place. Because these measures can cost a great deal of money, these precautions are often avoided or ignored in practice. Unfortunately, the cost of deficient safety measures is often passed on to the workers themselves when accidents occur. The economics of health and safety therefore involves weighing up the cost of safety measures to the contractor against the probability of accident and the severity of any accident likely to occur to the people on or near the site.

Part 2
The Economics of the Firm

7   Costs, Revenues and Pricing

Introduction

This chapter and Chapter 8 deal with costing within a firm. While Chapter 8 deals with the subject more from the point of view of traditional accountants, Chapter 7 tackles the subject more from the point of view of economists. Although accountants have always dealt with cash flow statements, showing money entering and leaving a firm over a period of time, in the past they were primarily concerned with auditing and producing sets of accounts such as profit and loss accounts and balance sheets. These accounts were essentially backward-looking, since they dealt with past events. Economists, on the other hand, are concerned with decision-making that concerns production in the future.

Although accountants are still employed to produce sets of accounts, since the advent of computing systems, the role of the accountant has changed and expanded. There is now an overlap between economists and accountants, since accountancy has become increasingly involved in management decision-making and management consultancy. Accountants are competent to deal with the figures supplied to them, but it is often up to others, such as production managers and economists, to provide the figures for sales, output and capacity. Management accounting is concerned with the economic implications of financial decision-making, and therefore the traditional approaches of economists and accountants have tended to merge.

Nevertheless, different questions require different answers, and in order to arrive at these answers, costs and revenues have to be calculated on a different basis. Accounts deal with questions concerning past performance, profitability and tax liability. Accountants deal with the problems of financial management, using cash flows and budgets to carry out project appraisals. These areas of accountancy require an exact assessment of the financial health of a company expressed in terms of invoices and receipts, and money flows.

Economists, on the other hand, deal with questions of viability, but their analysis is not bounded by actual or potential financial transactions. They wish to understand the behaviour of the firm in order to advise on its organisation, efficient levels of output, pricing policies and investment decisions. Moreover, many of the considerations that economists take into account depend on factors outside the firm itself. The rate of inflation, estimates of future demand, industrial relations and the degree of competition in a given market will all be factors that influence economic decision-making. Economists wish to know what difference a decision will make. They wish to know the best thing to do in a given situation, regardless of how the current circumstances might have arisen. The economists' approach is forward-looking, considers whatever options may be available and weighs these future options against each other.

Factors of production

As discussed in Chapter 4, the first thing any business needs to consider concerns its costs. A useful checklist for this purpose consists of the factors of production: namely, land, labour, capital and enterprise. They are the ingredients necessary for production to take place – the resources used up in the process of producing goods or providing a service. They are all, there-fore, costs of production.

All costs incurred by all firms and professional practices come under one or other of these headings: land may be used in the form of an office and the owners of land receive *rent*. Labour is not only provided by employees; partners may also contribute labour to a business, for which *wages* might be paid. Capital is often borrowed and the charge for it is called *interest*. If the capital has come from the partners' own savings, then interest payments should still be imputed, that is to say, as if interest had been paid to a bank. It is important to impute any costs, such as partners' wages, that are not actually paid. By imputing these costs it is possible to assess more accurately the real cost of providing the service. Otherwise, some of the resources used in the course of providing the service would be ignored.

All of the above costs have to be taken into account before *profit* can be calculated. Payments of rent, interest and wages usually involve a contractual arrangement, and the providers of these factors know in advance how much they will be paid. However, partners in a professional practice receive their returns in two forms: partly as wages or salary in return for hours of work, and partly as a share of the profits in return for their ownership of the practice. They can never be guaranteed a return for their inputs. Because private partnerships carry unlimited liability, losses may be incurred and even bank-ruptcy is a possibility. In limited companies the shareholders, who together own the firm, receive dividends out of profits. Table 7.1 summarises the relationship between factors of production and their respective money flows or returns.

Table 7.1   Factors of production

Factors of production	Returns to factors of production
Labour	Wages
Land	Rent
Capital	Interest
Enterprise	Profit

Rent

Rent is the amount charged for the use of a building or land. As it is of particular importance to those involved in construction and property, it is worth looking at in detail. More precisely, *ground rent* is charged for the use of the land. *Rack rent* applies to payments made in return for the use of buildings and forms the basis for leasehold agreements in England and Wales.

There is a third technical term used by economists, namely *economic rent*, which has a very specific meaning. David Ricardo, an early-nineteenth-century economist, coined the phrase to mean the amount paid for a factor of production over and above that necessary to keep it in its current use. Economic rent can be applied to all factors of production. If, for example, a worker is paid a wage of £250, but £150 would have been sufficient to attract and then keep that employee in the same employment, then £100 of that individual's wage is called economic rent.

In fact, Ricardo wanted to explain agricultural rents paid by farmers. Their rent depended on the fertility of the land they cultivated. As the land could not be put to any use other than agriculture, all rent paid for the land was economic rent. Landlords were in a position to increase their unearned income by charging more to the farmers of fertile than infertile ground, leaving the farmers equally well off regardless of the quality of the land they farmed. The extra or surplus produced by the fertile land compared to the least fertile land constituted different amounts of economic rent. The less fertile the land, the lower the economic rent that could be charged, although the land itself would remain in agricultural use. If any rent were charged on the least fertile land, the farmer would cease cultivating it and it would not be used. For the farmer, it would no longer be worth farming.

Karl Marx, who considered that there were only two factors of production – capital and labour – developed the idea of economic rent by describing three types of rent: *differential rent I*, *differential rent II* and *absolute rent*. Differential rent I is the additional rent paid because of variations in natural conditions, and differential rent II is the additional rent paid because of man-made improvements to the land, such as draining ditches and irrigation canals. Absolute rent is the rent paid for land, because the value of the output of each and every farm is greater than the cost of all the factors of production other than land. The difference between revenues and the cost of wages and materials is appropriated by the landlords in the form of rent. In fact, the landlords are in conflict with their tenant farmers because the higher the rent, the lower are profits. Thus, all farmers and firms pay rent. The lowest rent is equivalent to the absolute rent paid by all, while the differences in rent over and above absolute rent are accounted for by differences in natural fertility or man-made improvements.

Economic rent based on the economic or commercial productivity of land also applies to land in urban settings. Increasing the productivity of urban land by improving the use of sites increases the economic rent charged and this surplus is the source of developers' profits.

Several theories of urban land use account for the concentration of high rents in particular locations, including the theory devised by Heinrich von Thünen, whose 'rings', or contour lines, denote areas with particular levels of rent clustered round city centres, along main roads or in close proximity to facilities such as railway stations. As distance increases from those points of interest, with their economic advantages (such as proximity to markets and ease of communication), the use of land changes from largely commercial to largely residential and finally to agricultural uses. Rents vary, with the ability of building users to pay rent based on the nature and location of the site. This approach can be applied in principle to the rents firms and professional practices pay for their offices. The more prestigious the office, the more expensive the rent. Of course, other factors also play a part in determining the volume of work, the fees charged and profits earned by individual practices.

Profit

In the annual accounts prepared by accountants, profits are measured in terms of total revenues less costs. Economists, however, tend to take a broader view than accountants. To economists, the risk element should also be taken into account. Profit is the return to the owners, not only for their ownership of the business but also for their willingness to take risks. The greater the risk, the greater the anticipated profit needs to be, otherwise entrepreneurs would move their funds to safer investments offering similar profits. One reason for the high profits associated with high-risk projects is that even if the project does not completely fulfil expectations, there is an allowance that means the returns would still be positive. The riskier the venture, the greater this sum would need to be. If the project then succeeds, the allowance may not be needed and the risk-taker achieves a high rate of return.

The profits of a business should therefore be compared to the return that the same amount of capital could earn in an alternative use – its *opportunity cost*. For example, if the alternative is to invest the money in a bank account at 10 per cent per annum at no risk, then the profits should reflect the same return plus an element for the risk undertaken.

Profits also measure the financial success of an enterprise. These profits may be distributed to the partners, owners or shareholders, or they may be reinvested in the business. Without profits, shareholders would have no incentive to stay in the business. Profits are vital for investment. Investment funds come out of profits. In economics, profits can be *normal*, *supernormal* or *subnormal*.

A *normal profit* is the minimum profit required to keep a firm in business in the long run. Any less and the firm would withdraw from the market. A normal profit is equivalent to the amount of interest the capital would earn in, say, a bank account, plus a reasonable element for the additional business risk.

For example, two quantity surveyors set up a professional practice with £5000 capital, which they calculate will produce an annual return of £500 after all their expenses, including all labour charges, have been met. This represents a return on their investment of 10 per cent. If they could have earned 10 per cent on their capital without risk by depositing it in a bank, there would have been no profit from the practice, though it would still have provided them with a wage or salary. They must pay themselves a rate equal to the opportunity cost of their work. If they could have found work in another practice at a higher rate of pay than the wages they paid themselves, they would have lost potential income by working for themselves. Moreover, the cost of partners' time spent in the business but not actually paid should be imputed. This must be included to gain a true picture of the costs of the business. If a reasonable return on their capital after deducting their salaries, taking risk into account, had been, say, £1500, then £1500 would represent a normal profit. The return for taking the risk is £1000 and £500 is the amount the capital would have earned in any case. Any profits above normal profits are called *supernormal* profits.

An accountant may well report that a firm has generated a net profit, provided revenues are greater than costs. However, in the above example, if the profits had only come to £600, then profits would have been *subnormal*. Although the rate of return is greater than 10 per cent, it is insufficient to attract investment, given the risks involved. Investors might be willing to place funds in a project or a venture making subnormal profits or even losses in the short term, if they feel that profits in the longer term are likely to be normal, or even supernormal.

Actual losses are incurred when total costs exceed total revenue.

Costs

Total fixed costs (TFC) are those costs that do not alter with the amount of work carried out. For example, whether the practice is busy or not will not affect the rent due on the office premises. Fixed costs will depend on the terms of the contractual arrangement. At any moment the practice will be faced with commitments over which it has little control. In an architects' practice, for example, most costs will be fixed. These fixed costs include rent, community charges, secretarial salaries, bank interest and leasing costs.

Average fixed cost (AFC) is the fixed cost divided by the number of units of output. Average fixed cost is the fixed cost per unit. Output may be measured by the number of drawings, or the number of projects, or the value of fees. At low levels of output, AFC will be very high, as each drawing or project or £100 of fee revenue has to bear a high percentage of the burden of fixed costs. The rent, for example, has to be borne by a relatively small output. As output rises, the same total fixed cost is shared between more units of production. Each unit of output, as it were, has to carry less weight.

As can be seen from Figure 7.1, initially AFC falls rapidly. However, beyond a certain output, AFC will only fall very gradually as output increases. In other words, the AFC becomes less sensitive to changes in output as output increases. The average fixed cost makes it possible to find the level of output required to reduce to manageable proportions the contribution from each unit of output needed to cover fixed costs. In this case, if output is increased beyond Q_1, fixed costs per unit are not significantly reduced below C_1. Of course, the fixed costs themselves remain unchanged at C_2, and are therefore shown as a horizontal line in the diagram.

Total variable costs (TVC) are those costs which change when the amount of work changes. When output increases, some costs will rise. However, in practice, not all of these costs will be equally reversible. When extra staff are needed, the cost of recruiting is not equal to the cost of making them redundant again. Extra machinery may be bought new, but sold second-hand when no longer required.

Variable costs include materials, energy, labour and other costs, which vary directly with output: for example, temporary staff, overtime payments and hired plant. On a construction site, the more houses built, the more bricks,

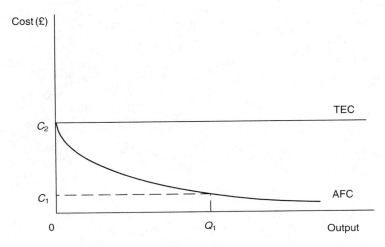

Figure 7.1   Fixed cost and average fixed cost

cement, windows and doors are needed. The total variable costs rise as output increases.

Average variable cost (AVC) is calculated by taking the total variable cost for a given output and dividing it by the number of units produced. In the above example, the AVC might be the cost of materials, plant hire and casual labour on the house building site. The TVC and AVC curves are illustrated in Figures 7.2 and 7.3.

The average variable cost is most useful in calculating the contribution of production towards fixed costs. Thus, provided that the price of a loaf of bread is greater than the cost of the ingredients, then the sale of that loaf will provide a small contribution towards fixed costs. As long as there are net contributions to fixed costs like rent, salaries and interest payments, then production will have helped to reduce these liabilities of the firm, even if the firm is trading at a loss. This last point is most important. It suggests that although a firm may be making losses, there are circumstances in which the firm should continue in business, at least in the short term.

Total cost (TC) is the sum of all fixed and all variable costs for a given level of output. Figure 7.4 illustrates the relationship between fixed, variable and total costs.

Each level of output has, therefore, to be calculated separately. Again, the actual measurement of output is not as important as finding a consistent unit of measurement. Once that has been achieved, then total costs may be associated with different levels of output.

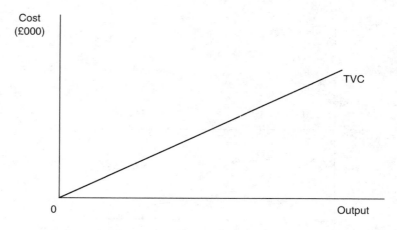

Figure 7.2   The total variable cost curve

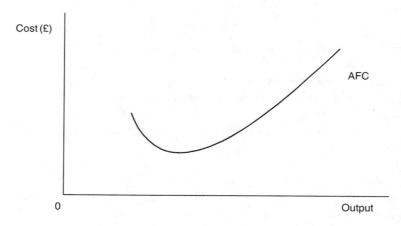

Figure 7.3   The average variable cost curve

Average total costs or *average costs* (AC) refer to the total cost of each unit produced, taking fixed as well as variable costs into account. The period of time covered by costs is important. In the short run some costs are fixed, but in the long run all costs can be varied. The length of the short term depends on the length of time of the longest commitment. Thus, if a firm's longest commitment or agreement is to pay rent for two years, then two years is the short-term period for planning purposes. In two years the firm may decide to expand into larger premises and therefore increase its rent payments. The short term will vary from firm to firm and from industry to industry.

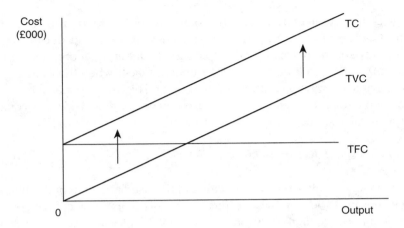

Figure 7.4   *Fixed, variable and total cost curves*

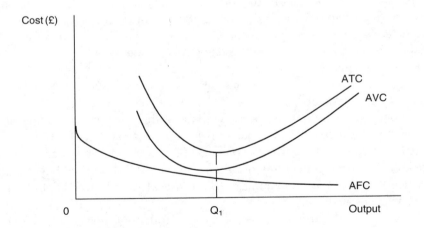

Figure 7.5   *Average fixed, average variable and average total cost curves*

In the short-term, therefore, total costs may be divided between fixed costs and variable costs. The short-run average cost, or the cost per unit of output, can be derived from the average fixed and average variable costs by combining the vertical heights of the AFC and AVC curves. As these costs depend on the quantity produced in a given period – say, per day or per month – they may be plotted against output as in Figure 7.5, which shows the short-run average total cost curve.

It is important to establish the total cost of production for each unit produced in order to monitor the efficiency of firms and decide on the methods and techniques used to produce a certain quantity of output. In Figure 7.5, the firm is at its technically most efficient when the average total cost is at a minimum at output Q_1. At lower and higher outputs, each unit will cost more to produce, reflecting the use of more (or more expensive) human or material resources for each item produced.

Economies and diseconomies of scale

Economies of scale are savings because of the size of the firm or the quantity of its output. When a firm can reduce unit costs simply by increasing output, it is in a position to take advantage of economies of scale. These savings may be caused by internal or external factors.

Internal economies of scale may result from better use of expertise and specialist knowledge within a firm. Large orders may save on administrative costs. One order to a manufacturer for 1000 units takes the same amount of time to write as an order for 10, reducing administrative overheads per item. Machinery such as tower cranes can be justified only if the volume of work enables the equipment to be operated at or near its designed capacity.

External economies arise out of the concentration of similar firms in a locality. Each firm is then able to recruit new staff more cheaply. Suppliers may also locate nearby for convenience. Firms will have closer access to developments in the market, possibly enabling them to react to changing circumstances more efficiently than outlying businesses. External economies of scale are reasons for potteries congregating in Stoke-on-Trent, banks and stockbroking firms concentrating in the City of London, and architectural practices locating in London and the South East, where just over 45 per cent of all practices are based.

In Figure 7.6, as output increases up to an output of Q_1 units, the cost per unit declines, reflecting economies of scale. These economies arise because overhead costs such as management and administration costs do not rise as fast as output. Moreover, as output increases, better use may be made of equipment the more it is utilised. However, beyond Q_1, the cost per unit begins to escalate as overtime working or added incentives have to be paid to complete work. A firm can therefore take advantage of economies of scale up to a certain level of output, but if it continues to expand beyond that level, with the same number of employees, same size of offices and same equipment, then diseconomies of scale will be incurred.

If diseconomies are present in the operation of a firm, then it can raise its efficiency simply by reducing its level of output. This will have the effect of

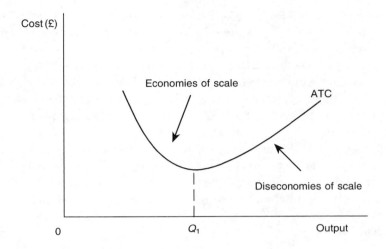

Figure 7.6   *Average total cost curve, showing the optimum level of output and economies and diseconomies of scale*

reducing its cost per unit of output and help to make it more competitive. Thus, Q_1 units per month represents the optimum level of output in this example.

In manufacturing industry, *mass production* enables firms to take advantage of economies of scale. Research and development costs are then spread over many units of output, reducing the cost per item.

Mass production techniques are also applied in the construction industry. These techniques mainly take place off-site in factories where standardised prefabricated components are produced on a repetitive basis. The manufacture of building components prior to arrival on site includes standard fittings for kitchen units, toilet units and systems of fenestration. Curtain walling and pre-cast concrete sections of buildings require either simple assembly on a particular site or specialised assembly by the manufacturer's own trained gang. Standard floor to floor heights and floor to ceiling heights in public sector buildings have also reduced some component costs, such as staircases and lifts. Improved techniques have also been applied to the mass production of materials such as timber, metal, and especially brick. As a result, the cost of these materials has not risen as rapidly as the cost of labour or land. However, the unique nature of every project, the ground problems of different sites, and the particular requirements of individual clients often in practice reduce the scope for mass production on site. Prefabricated housing was used extensively before and after the Second World War, up to around 1955. The main types of prefabricated housing were concrete, steel-framed or timber. There are one or two examples of system building, especially schools and factory units, which continue to be supplied in ready-made units.

Prefabricated components help to cut building costs because they simplify the construction process on site, reducing site labour costs and site management. In fact, the significance of savings related to prefabrication in construction depends on the percentage of total costs represented by manufactured components and the likelihood of continuous production runs. Total construction costs also include the cost of land, site labour, plant hire, and professional fees. Since the Second World War there has been an increasing use of prefabricated components as a proportion of total costs. Materials and components have become cheaper (relatively) than land and labour.

Marginal cost

Marginal cost (MC) is the increase in total cost resulting from producing one more unit. Alternatively, MC is the reduction in total cost that would occur if the last unit had not been produced. Thus, MC is the cost associated with the extra unit or the last unit produced. In construction, firms produce work for projects, and each project may be seen as a unit of work. In construction, MC can be used to refer to the increase in total cost resulting from taking on one more job or the amount saved by not taking it on. It is the additional cost incurred in order to take on extra work.

A firm of electrical engineers would not necessarily need to buy a new van, purchase extra equipment or hire additional labour when emergency work had to be carried out during normal working hours if these were available in the firm. The marginal cost of doing the job would be the cost of fuel, wear and tear on the vehicle used, materials used in carrying out repairs, and perhaps some telephone calls and stationery. The MC of that job is far less than the total cost, TC, which includes the cost of labour, a proportion of the cost of the van and equipment used, and a contribution towards the overhead costs of running the firm.

A simple formula for MC is:

$$MC_n = TC_n - TC_{n-1} \tag{7.1}$$

where

n = number produced
MC_n = marginal cost of the nth unit
TC_n = total cost of producing n units
TC_{n-1} = total cost of producing 1 less than n units

Applying this formula to a firm of consulting engineers, assume that the total costs of providing the services of the firm were £500 000 over a period of one year. If the total cost, omitting the cost of undertaking the last project would have been only £420 000, then the MC of the last job is £80 000.

The relationship between MC and AC can be seen in Figure 7.7. When the MC is below the AC curve, the AC curve is declining. When the MC is above the AC, the AC curve rises. This relationship is easily explained. Imagine a class of students sits a test. Each student is given a mark. The class average is 60 per cent. If a new student takes the test and scores below 60, the class average will go down, but if the mark is greater than 60, the class average will rise. Only if the new student scores 60 per cent will the class average remain the same. In Figure 7.7, the MC equals AC at the minimum point on the AC curve at, say, Q_1 units per month, which is, as noted in Figure 7.6, the most efficient level of output.

It is often impossible to isolate the marginal cost of the last unit. For example, bricks are delivered in batches. The *incremental cost* is therefore used to calculate the increase in total cost as a result of ordering one more batch. An estimate of the marginal cost may then be derived from the incremental cost by dividing the incremental cost by the size of the batch. Assume that bricks can only be bought in batches of 1000. Thus, if the incremental cost of an additional batch of 1000 bricks is £1000, then the MC of a brick is £1. If 1500 bricks are required and a second batch is ordered, the incremental cost of the second batch is still £1000, but the marginal cost of a

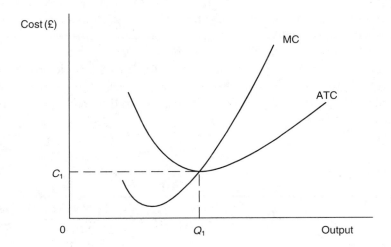

Figure 7.7   Marginal and average cost curves

brick in the second batch is £2. The remaining 500 bricks are surplus to requirements, but still need to be purchased.

Other costs

In Chapter 1, it was noted that the central problem of economics is the fact that resources are scarce. There is never money enough to do everything we would like. Therefore, it is always necessary to give something up in order to do or to have something else. The real nature of cost is more than just the money paid. The money price only enables exchange or transactions to take place. The price of a cup of tea tells us its exchange value but little about how the individual feels, or what else might have been purchased instead.

The *opportunity cost* is the opportunity lost. It is the cost of doing something in terms of the next best alternative. When a building is erected on a site, the opportunity cost is the next best alternative use of that site. When one pays for something, one is sacrificing what that money could have bought instead. A cost is therefore a sacrifice. A £10 meal in a restaurant would cost a wealthy industrialist less than it would a poor student. While the industrialist may not have to give up anything, except perhaps savings of £10, the student may be forced to give up, say, the purchase of a textbook.

This is a much broader definition of cost than simply the money value, and is critical in making management decisions. The importance of opportunity costs for management, both within the office and on construction projects, is that alternatives must always be considered in full before a decision is made. Opportunity costing lies at the heart of economic theory and method. Assume, for example, a firm with a budget of £100 000 has a choice between two offices, as illustrated in Table 7.2. If Option A costs £70 000 and Option B £80 000, then choosing the cheaper option may not necessarily be the correct decision. The opportunity cost of Option A includes the forgone expected profits of Option B. Thus, if the firm chooses Option A, it gives up profits of £120 000, which is more expensive than the sacrifice of £30 000 if Option B is selected.

Table 7.2   Opportunity cost

	Option A (£)	Option B (£)
Cost of office	70 000	80 000
Expected profits	100 000	200 000
Net expected profits	30 000	120 000

Sunk costs are those costs that were paid in the past and cannot be altered. They have no opportunity cost because they have no current alternative use. They do not affect any decisions that need to be taken currently because, whatever the decision, the money spent cannot be regained.

The concept of sunk costs serves as a reminder that it is the future consequences of decisions that matter. Any mistakes made in the past are just that. They are past errors precisely because they hinder progress now. It is undesirable to spend time in the present situation trying to put past matters to rights instead of advancing to an improvement. Let bygones be bygones.

It would be an error to consider sunk costs when making management decisions. A simple example will illustrate the position. I leave my house to walk to my local bus stop. I turn right instead of left. After 100 metres I realise my mistake. I do not continue walking in the wrong direction in order to justify my decision, but turn around immediately and walk to the bus stop.

Occasionally, time is spent at the planning stages on projects which fail for one reason or another to get on site. The decision to pursue unlikely projects cannot be based on the amounts of time, money and effort that have been spent on them up to that point. The pertinent question to ask is: would more time on a given scheme be more likely to provide a return than time spent on another project? Decisions should be based on the best thing to do in the current situation with reference to the present and the future only. From a firm's point of view, it is a mistake to correct past errors at the expense of potential future income.

An *imputed cost* is a cost for which no invoice is rendered but which none the less represents an amount of a resource that has been used up. An imputed cost in a surveyor's office may well be the extra time spent after hours by partners for which they receive no payment as such from the practice but is nevertheless time which may be attributed to one project or another, or indeed to running the practice as a whole. Such imputed costs do not affect the cash flow of the practice, but need to be taken into account to establish the economic viability of a project or to ensure the true profitability of a contract.

Having discussed the implications of various cost concepts, we now turn to the various types of revenues. If costs represent money leaving the firm, revenues represent cash flows entering it.

Revenues

Total revenue (TR) is the total income of the firm. It is found by multiplying the quantity sold by the unit price. Thus

$$TR = P \times Q \tag{7.2}$$

where

P = price
Q = quantity sold.

This formula applies where identical products are sold. In construction, each project is unique. TR is therefore calculated by adding the revenue from each project in turn. Total revenue not only relates to sales or fees for construction services, but also includes income from sublet property, perhaps even royalties and shares of profits from developments, which may have been carried out by the firm. TR is the total of all sums from all sources. Nevertheless, returning to the general model in which the firm sells only one product line, the TR curve begins at the origin, rises to a maximum level and then declines. This can be seen in Figure 7.8.

Average revenue (AR) is calculated by dividing the total sales revenue by the output, or the number of units produced. The AR in an architectural practice is the total fee revenue divided by the number of projects undertaken, or it might even be calculated as the total fee revenue divided by the number of completed drawings, depending on the nature of the question. For example, there may be a need to know the value of project work on average,

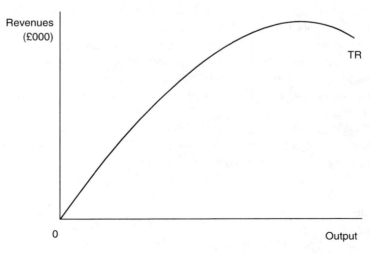

Figure 7.8   Total revenue curve

or a firm may wish to charge for drawings but needs to know the average price it has charged in the past. The resulting AR can then be used to compare performance from one year to another, but, of course, as with all averaging, it totally ignores the uniqueness of each project and the variation of effort required for each drawing.

The general shape of the AR curve is shown in Figure 7.9, which illustrates that as output increases, the revenue per unit declines. This reflects the conventional argument that in order to sell more, the price (and hence revenue per item) must be reduced. The average revenue can be applied to predict total revenue at different levels of output. This is done by plotting the AR curve on the basis of past combinations of price and corresponding quantities sold. Once the AR curve is known, TR is found by multiplying AR (or price) by a given level of output. In Figure 7.9, if Q_1 units are sold at a price per unit of P_1, then total revenue is equal to $Q_1 \times P_1$.

Marginal revenue

The *marginal revenue* (MR) is the increase in the total revenue resulting from producing one more unit. If the unit of production is a project, then the marginal revenue is the difference the last project makes to total revenue. If total revenue increases by the amount paid for the last project undertaken, the fee or the contract price is equal to the marginal revenue.

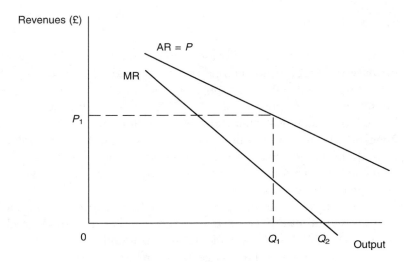

Figure 7.9   Average and marginal revenue curves

However, this does not always occur. Sometimes it is necessary to lower all prices in order to sell one more unit. The revenue from the last unit then has to be set against the reductions made on all earlier units sold. For example, assume that a speculative builder has a site with planning potential for ten or eleven houses. Assuming ten houses are built, they could be sold for an expected £100 000 each, and if eleven houses are built, their unit price would need to be reduced to £95 000.

The total revenue of 10 houses	10 × £100 000 =	£1 000 000
The total revenue of 11 houses	11 × £95 000 =	£1 045 000
The marginal revenue of the 11th house	=	£45 000

Thus, although the selling price of eleven houses is £95 000 per unit, the marginal revenue of the last house is only £45 000.

The general formula for MR may be calculated as follows:

$$MR_n = TR_n - TR_{n-1}$$

where

$$n = \text{number sold}$$
$$MR_n = \text{marginal revenue of the nth unit}$$
$$TR_n = \text{total revenue from selling n units}$$
$$TR_{n-1} = \text{total revenue from selling 1 less than } n \text{ units}$$

Profit maximisation

In Chapter 4, the aims of firms were discussed. One of the main aims was seen as the maximisation of profits, although it was not made clear whether profits should be maximised in the short run or the long run. Even if short-run profit maximisation is not the primary aim of firms, it is useful to study it in order to understand the theoretical interactions between the various cost and revenue concepts. The relationship between costs and revenues is illustrated in Figure 7.10. The vertical axes of parts (i) and (iii) show costs and revenues measured in £m, representing large amounts, whereas part (ii) is measured in £, signifying much smaller unit prices.

Figure 7.10 (i) relates TR to TC. Profits are equal to total revenue minus total cost. At an output level of Q_2 units the difference between the TR and TC curves is greatest, $TR_1 - TC_1$, and profits are maximised.

The marginal revenue is the change in total revenue as output increases, and can be seen as the gradient of the total revenue curve. Similarly, marginal

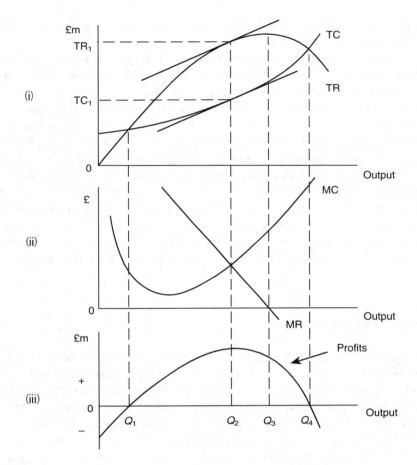

Figure 7.10   The derivation of profits

cost is the gradient of the total cost curve. In Figure 7.10 (i) at the profit-maximising level of output, the gradient of the TR curve and the slope of the TC curve are equal at an output of Q_2 units, shown by the two parallel tangents. Thus, MC = MR, as shown in Figure 7.10 (ii), where the MC curve intersects the MR curve at Q_2 units.

It can be seen in Figure 7.10 (i) that total revenues at TR_1 are not at a maximum at the profit-maximising level of output. To maximise revenues would involve raising output to Q_3 units but this would increase costs more than revenues, and therefore reduce profits. This is confirmed by looking at the profits curve in Figure 7.10 (iii). This shows profits rising to a peak at Q_2 units and then declining at Q_3 units.

In Figure 7.10 (ii), the MR curve crosses the x-axis at Q_3. At this point, the MR equals zero and this is confirmed in Figure 7.10 (i), as TR does not increase or decrease by the addition of the last unit of output.

Finally, in Figure 7.10 (i), it can be seen that between zero and Q_1 units of output, TC is greater than TR. As a result, losses are shown in Figure 7.10 (iii), where the profits curve is below the horizontal axis. Output at Q_1 is the lower break-even point. Between Q_1 and Q_4, TR is greater than TC but at output above Q_4, the upper break-even point, losses are again incurred. It is usually preferable to be small and profitable than large and loss-making.

From this analysis of costs and revenues, a simple strategy emerges for firms involved in construction projects. To increase profits, the increase in total revenue must be greater than the increase in total costs. Thus, a firm should take on additional work as long as the MC of extra work is less than its MR.

Marginal cost pricing

Marginal cost pricing is a pricing technique, based on marginal costs – which are the additional costs of carrying out a particular job, bearing in mind that fixed costs would be incurred in any case. Therefore, marginal costing takes into account the increase in variable costs resulting from the last unit pro-duced. If the anticipated revenue is greater than the marginal cost, then the difference is a contribution towards overhead costs, such as rent, interest payments and other head office costs. If, on the other hand, the marginal cost of a particular project is in excess of the marginal revenue, the firm must cease work as soon as it can. Otherwise, profits are, in effect, being used to subsidise some clients at the expense of the others.

When firms tender for work, the concept of marginal revenue is most useful. It is essential that firms price competitively, which means they must be aware of the minimum price they are willing to charge (which is the marginal cost). Marginal cost pricing is also used when firms are desperate for work and need to generate a cash flow. The marginal cost ignores fixed costs and focuses on changes in variable costs only. Any revenue in excess of the variable cost of a job can then be used as a contribution towards fixed costs. In simple terms, instead of having to find the full rent, at least part of the rent is paid. It would be in the firm's financial interest to accept such work. Marginal cost pricing indicates the financial viability of a job, provided that the fee charged to clients is greater than the extra cost.

Cost plus pricing

The usual method of costing and pricing work in construction is *full cost pricing* or *cost plus pricing*. Cost plus pricing involves calculating the variable costs of labour, materials and plant hire related to a project. A proportion of fixed costs is then added, depending on how large a percentage of turnover a particular project is likely to be. A percentage mark-up is then added to the total cost to determine the final price to be charged or tendered. If this method is adopted, the minimum price would need to cover both variable costs and a proportion of fixed costs, plus an element for profit, compared to the minimum price of the marginal costing method which only requires that variable costs are met.

The problem with cost plus pricing is that in the course of a year it is not possible to know the percentage necessary to ensure that all costs will have been met by the end of the year. If the percentage mark-up is too low, fees will be insufficient to cover all costs, and this can only be known with the benefit of hindsight. On the other hand, if the percentage mark-up is too high, the firm will be less competitive and, as a result, lose bids in the tendering process it might otherwise have won.

Forecasting

Governments and firms attempt to forecast construction demand. Forecasts of building demand are needed by the government to monitor construction and to plan adequate supplies of materials and components, as well as trained labour. If construction demand increases unexpectedly, any shortfall in supplies tends to lead to inflationary price increases or to be made up by importing materials and components from abroad.

Whatever business strategies contractors adopt, they need to forecast demand in order to anticipate the amount, location and type of work likely to become available. Firms which supply construction services, design and management, components and materials, and hired plant and equipment, also need to predict demand for their products and services. Just as the demand for contractors is a derived demand, the demand for building components is a derived demand dependent on the demand for construction work. For example, the greater the amount of expected construction work, the more materials will be required.

Derived demand forecasts for materials are based on resource coefficients per £1000 of construction output. A resource coefficient is a number or fraction such that if the resource coefficient of a material, say cement, is 0.05, then for every £1000 of anticipated construction demand, demand for cement would be £1000 multiplied by 0.05, namely £50. If expected construction demand was £10m, then demand for the material would be £0.5m. This figure could then be compared to existing stock, production plans, and current sales.

A variety of academic institutions, government departments and private consultancy organisations regularly undertake predictions of construction demand. These forecasters include the Treasury, the National Institute for Social and Economic Research, Cambridge University, Oxford University, the London Business School, the London School of Economics, the National Council of Building Materials Producers, and several other organisations with an interest in the construction and property sectors.

In order to make their predictions, use is made of economic indicators. Some indicators are known as lead indicators, showing trends in advance of a rise in construction activity. Other statistics are lag indicators, which reflect changes in construction work. The following are examples of construction economic indicators.

Applications for planning permission

Applications for planning permission are also an advance indicator of construction demand, although many applications are speculative, and many others do not come to fruition. These figures are kept by local authorities.

Land prices

Land prices might be a lead or lag indicator of construction demand. Most speculative developers own a land bank of 5–6 years of supply for houses. They buy land in slumps when it is cheap, and run down their stock in booms, when land is costly. House builders speculate contra-cyclically. Often they are subsidiaries of financial institutions or large builders and can therefore afford to hold land. Once the rate of land price increases begins to slow down, the pressure of demand for future building work can be expected to follow. Similarly, when the rate of decline in land prices begins to slow down, pressure of demand for construction can be expected to grow, after a time lag. Data on land prices can be found in journals such as the *Estates Gazette*.

House builders' register of starts

The house builders' register of starts for insurance purposes is kept by the NHBC. This shows house building trends. There is a time lag between starting and completing dwellings. Timber-frame construction reduces the construction period. The start-to-completion time of a whole site can vary, because builders use batch production techniques, which can be varied to match demand or sales. Statistical trends can be perceived using a time series of starts.

Other indicators of construction demand

Firms' invitations to tender can be seen in publications. For example, the *Contract Journal* lists jobs by type and by region. Local authorities' and other large organisations' plans, where published, also indicate potential changes in construction demand in particular areas. ICI on Teesside, regional health authorities and British Coal in South Wales have indicated periodically or implied through public statements, their expected future building requirements.

Other methods of predicting construction demand are based on the economy as a whole. These methods use what economists call macroeconomic indicators. For example, if construction demand depends on the size and rate of growth of GDP and if construction output is 6.5 per cent of GDP, then an economic model of total construction demand could be based on a prediction of the GDP. In other words, construction demand is equal to the GDP multiplied by 6.5 per cent. More complicated economic models of construction demand can take interest rates, unemployment and other economic variables into account.

It should be noted, however, that the effect of interest rates and unemployment on construction will tend to vary from region to region, because, as we have seen above, markets vary because of different local factors. For example, some areas experience higher levels of unemployment than others. Industries and commercial activities are not spread evenly throughout the country. Demographic changes are constantly taking place as people migrate, especially in search of work. Changes in the South East area of England often lead to changes in other areas following a time lag.

It is simply not possible to predict the long term future with any accuracy. There are too many unknowns: wars, new techniques, changes in tastes or living patterns, climatic changes, government policies, environmental impacts and solutions, and rising costs of materials, land and labour. Nevertheless, several commentators have predicted that over the next fifty years there will be less new build and more work on existing stock. However, such predictions must allow for wide regional variations as well as variable rates of change from one year to the next. Although it is clearly difficult to rely on predictions, the process of discussing trends relating to the above list of variables can still be useful in assessing current policies and for making assumptions for planning purposes.

8 Markets

Introduction

As far as marketing is concerned, compared to those selling other products and services, the construction industry and building professionals are faced with a particular set of problems. Carpenter (1982) defines these difficulties as follows:

> The purchaser generally doesn't know what he wants when he starts to buy it, no-one can actually be sure that what he requires can be produced, the production capacity to produce it doesn't exist at the time of commissioning and there are a large number of bodies and officials whose job it is to stop you getting what you want. . . . In summary the building industry and its professionals sell a production service, not buildings.

This chapter develops the context of the marketing role by examining different market types. This has a direct bearing on the tactics used in the marketplace and helps to show the likely reaction of competitors to any moves a firm may make.

Perfect competition

Some markets are more competitive than others. The most competitive markets, which economists describe as *perfectly competitive*, are those in which all of the following conditions must necessarily be found.

Product or service homogeneity

The product or service offered by each firm must be identical to the product or service offered by every other firm. However, firms like to point out how they differ from their competitors. Indeed, much time is spent showing potential clients the subtle differences between the services offered, to point out the benefits of one firm over another. This marketing process is known as *product or service differentiation*. Moreover, in the construction sector, all projects are unique. The product or service is therefore not homogeneous.

Free entry and exit

There must be free entry and exit to and from a market. Because it requires little capital to set up in business in the construction industry, there are no financial barriers to entry as such. It is not essential for a firm to purchase its own equipment, which can be hired. Labour is also hired when needed, and materials are often obtained on credit. However, even in construction, there are several hidden barriers. Personal contact is often essential, as trust is important in building projects. Short-lists of professional firms and contractors are often drawn up on the basis of personal contacts over a long period and past experience of working together. New firms therefore often need time before they can enter into competition with established companies or professional practices.

Many buyers and sellers

There must be many buyers and sellers, and there is no loyalty between buyers and sellers. This enables buyers to find alternative suppliers, while sellers are not need to sell to any one particular customer. Everybody in a perfectly competitive market is a price taker. There is a going rate for the job and no one would be able to undercut or sell above the market price. Whatever any one firm decides to do, it will make no measurable difference to the market price, because in perfect competition it only produces an extremely small fraction of the total supply. It can sell all its output at the going market rate.

Perfect knowledge

Perfect competition assumes perfect knowledge of prices, costs, technology and the competition. Firms wishing to enter the market and clients wishing to place orders will therefore have access to all the information they need in order to make decisions. Without that knowledge it is possible for a firm to take advantage of its customers, suppliers or competitors, setting a price above or below the going rate.

All firms have trade secrets and price-sensitive information. They do not share their client lists, sources of supply, or detailed costs. The assumption of perfect knowledge is simply not achievable. A market is competitive only to the extent that it complies with these conditions. In short, perfect competition does not exist. It is purely a set of theoretical rules.

Imperfect competition

Perfect competition is a theoretical set of conditions, which acts as a yardstick against which the competitiveness of any given market can be compared. Whenever any of the conditions are broken the market is said to be imperfectly competitive. According to neoclassical theory, imperfections in markets enable firms to raise their selling prices above what would have occurred in strictly competitive situations. This can be seen in Figure 8.1.

In perfectly competitive markets, firms would be forced to produce at their most competitive level of output where their costs per unit were at a minimum at Y on the AC curve. In Figure 9.1, this would occur at output Q_2, where costs are only C_1. However, in imperfect competition, firms will produce at Q_1, where MC = MR at Point X, assuming they wish to maximise their profits. At this level of output, the cost per unit rises to C_2. In imperfect competition, firms produce at less than their most efficient level of output, and costs per unit are above the lowest possible cost. Nevertheless, companies are still in a position to make supernormal profits. At a level of output of Q_1, firms can charge what the market will bear at Z, namely a price of P_1.

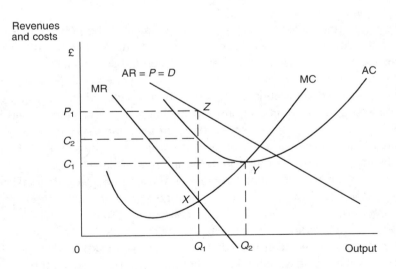

Figure 8.1   Imperfect competition

Monopoly

In theory, a monopoly is defined as a single seller in a market. This gives the monopolist a great deal of power and influence over buyers, because the single seller can choose either the price or the quantity of the product to be offered for sale. The actual degree of monopoly power depends on the availability of substitutes.

By restricting output and creating artificial shortages, monopolists can raise prices charged, thus increasing revenues while lowering output. This will enable the monopolist to maintain supernormal profits even in the long run. As this has been seen by many to be against the public interest, several Acts of Parliament have been introduced since the Second World War to control monopoly power. Monopolies legislation includes the Monopolies and Restrictive Practices Act of 1948, which set up the Monopolies and Mergers Commission. The Fair Trading Act of 1973, which set up the Office of Fair Trading, also defined a monopoly. A firm which controls at least 25 per cent of the market in a given product or service or, alternatively, a group resulting from a merger with gross assets of at least £5m, can be referred to the Monopolies Commission under the legislation. In 1984 the value of assets was raised to £30m. Both the Monopolies Commission and the Office of Fair Trading are intended to protect the public interest, which was defined in the 1948 Act in terms of efficient production and the 'best distribution of men, materials and industrial capacity'.

Although monopolies are capable of acting against the public interest, it does not follow that they are necessarily a bad thing. Research is often too expensive and too long-term for small firms to undertake. Economies of scale may make it cheaper to allow a monopoly to continue trading. Monopolies may help to avoid duplication of services. Thus it is preferable to have one gas company supplying an area in order to minimise the disruption caused by pipe-laying, or one interconnecting national telephone system to give all telephone users access to each other. Firms which are a monopoly at home may be just one of many firms competing internationally, where competition is between large, multinational corporations. Without an adequate market at home, especially where that market is not large enough to support more than one large firm, it would not be able to compete internationally. Finally, the take-over of weaker firms by more efficiently run companies means the most efficient firms survive and grow to dominate their market. Restricting successful firms from merging with or taking over other firms may prevent them from improving their efficiency, with the result that they are less able to compete against competition from abroad than they otherwise would be.

Oligopoly

The most common type of market competition is *oligopoly*. An oligopoly is a market dominated by a few sellers. High street banking is dominated, among others, by National Westminster, Barclays, Lloyds TSB Group, Midland and the Royal Bank of Scotland as well as the Co-operative Bank. The oil industry is concentrated in firms such as Shell, British Petroleum, Texaco, Jet and Esso. Even grocery retailing is oligopolistic, dominated by a few firms such as Tesco, Sainsbury's, Safeway, Asda, Waitrose, Marks & Spencer, and one or two regional multiples. Domestic airlines, travel agencies, insurance, building societies, vehicles and alcohol are all examples of oligopolistic markets.

Although there are many thousands of firms in construction, the industry is fragmented into trades and regions. Thus, in any one region there are only a few firms that dominate in any particular market. For example, in 1991 there were 207 400 private contractors in the UK, of which 14 943 were based in the East Midlands. Of these, only six firms employed more than 600 people. Taking the country as a whole, firms employing 600 or more people undertake almost 25 per cent of all work. If this is reflected in the market in the East Midlands, only six firms undertook almost a quarter of construction work in the region.

Any rise in demand for an oligopolist's output is derived from its share of a growing market or at the expense of its competitors share. However, firms in oligopolistic competition face a dilemma. They can lose revenues if they raise or lower their prices. If a firm raises its price, other firms will tend to hold their prices constant. If the firm reduces its price, the others will lower theirs similarly. Nevertheless, in oligopolistic markets there is the constant threat of a price war, when firms attempt to undercut each other, even selling at a loss, to gain or defend a percentage point in their share of the market. Often, an uneasy peace is maintained by tacit agreements between sellers in order to avoid a damaging confrontation. To avoid such conflicts, firms on occasion organise themselves into a cartel or group of firms acting in concert with each other.

Agreements between firms in oligopolistic competition may be drawn up in order to maintain an orderly market. For example, the airline industry has maintained that agreements between airlines are necessary for public safety. Where agreements are explicit, they must be made public. Occasionally they may be seen as being against the public interest. Under anti-monopoly legislation, they can be referred to the Office of Fair Trading. However, restrictive practices and secret agreements are by their nature difficult to legislate against. Some years ago, collusion occurred between some members of the ready-mixed cement industry. Having met secretly, they would tender in such a way that one member would price a job at slightly less than the other members of the cartel. Had they been genuinely competing with each other, a much lower price would have been obtained. These secret agreements enabled all members of the cartel to take their turn, each charging above competitive market prices.

Oligopolistic arrangements designed to maintain high prices and restrict competition between the dominating firms in a market do not always protect firms as much as would first appear. The high prices cushion inefficient producers and allow costs to rise above a competitive level. In spite of their relatively high costs, oligopolistic firms can remain profitable. By restricting competition, firms do not necessarily feel the need to diversify or expand into new markets, but concentrate their efforts on protecting their positions. When new products and technologies emerge abroad, oligopolists are frequently at a competitive disadvantage internationally. Maintaining barriers to entry make oligopolists dependent on these barriers for their survival. In neoclassical economic terms, when the barriers are reduced, then structural changes in industries occur more rapidly than might have been the case had competition allowed firms to merge or leave the market over a period.

Contestable markets

The above theories of competition do not always account for the behaviour of firms in markets, especially if prices are not raised and output is not restricted, even where competition is minimal. In attempting to understand the behaviour of firms, the concept of contestability developed by William Baumol and others may be helpful.

In a contestable market, firms act *as if* it were highly competitive because they fear the arrival of new entrants. This fear would induce them to maintain standards of service, or the quality of the product at competitive prices, to discourage extra competition and maintain their current share of their market.

Contestability is a measure of the threat of new competition to existing firms, where barriers to entry are relatively weak. A high price would give new outside firms an opportunity to undercut existing firms. Deliberately restricting output would immediately give potential competitors a chance to enter a market, if they were able to produce the shortfall. The concept of contestable markets shifts the emphasis from the number of sellers in a market to the ease of entry of new firms into it. The assumptions of a perfectly contestable market compared to those for perfect competition can be seen in Table 8.1.

Table 8.1   Comparison of competitive and contestable market conditions

Necessary assumptions	Competitive market	Contestable market
Profit maximisation	Yes	Yes
No barriers to entry and exit	Yes	Yes
Perfect mobility of inputs	Yes	Yes
Perfect information for all traders	Yes	Yes
Large number of small firms acting as price takers	Yes	Not required
Homogeneous product	Yes	Not required
All firms or (potential firms) are faced with the same cost functions	Yes	Yes
Average cost curves are U-shaped	Yes	Not required

Source: McDonald, 1987, p. 183.

This table illustrates that fewer assumptions are required for a contestable market than a perfectly competitive market. For example, a perfectly competitive market requires all firms to be price takers, whereas, even if there are only a few incumbent firms in a contestable market, they will be aware that potential competitors will be attracted into the market if prices are seen to be high and very profitable. The threat of new competitors arriving on the scene will be sufficient to keep prices in the market at a competitive level. However, Schwartz (1986) maintains that the ability of firms to drop prices rapidly will offset this threat. Markets, he argues, remain, 'uncontestable in the sense that price behaviour becomes unaffected by the threat of entry. Available evidence indicates that this is typically the case'.

According to the theory of contestable markets, the fear of new entrants and increased competition may deter some firms from charging more than they do, even when their experience, expertise or reputation might appear to give them an advantage over other practices. Nevertheless, as Table 8.1 makes clear, the assumptions needed for the theories of competitive and contestable markets are completely unrealistic.

Alternative approaches to understanding markets

Markets are a method of resolving disputes between people. Buyers and sellers must agree a price, but to a seller a good price is a high price, and to a buyer a good price is a low price. Markets allocate goods and services, but the more goods allocated to one group, the less goes to another. Conflict is inherent in society, and markets need to be seen in the context of the social relations that determine the bargaining strengths of the buyers and sellers.

As we noted earlier, effective demand is the quantity of goods people are willing and able to purchase, but what determines ability to pay? Income varies, and those on low incomes are simply excluded from markets, while those with income or wealth can afford to continue buying. For example, when food prices rise, the poorest are the first to be affected. Similarly, house prices are so high that many people simply cannot afford adequate housing. Raising house prices does not solve problems of homelessness. In labour markets, especially markets for unskilled labour, many people compete for work. There are usually unemployed people available to take over from any-one who is unwilling to work for a given rate of pay, or who leaves a job. The competition for work strengthens the bargaining position of employers. Even skilled workers can find their job security, conditions of employment and wages undermined by changing technology, which can weaken as well as strengthening their bargaining position. Markets need to be seen in their social and economic contexts.

One of the problems of conventional economics is that markets are treated as if they were abstract concepts based on sets of assumptions. In fact, they are sets of rules agreed by, or imposed upon the participants. In some markets, firms do behave in patterns similar to imperfect competition, oligopoly or monopoly. However, Morishima (1984) has described three types of trading between buyers and sellers which distinguish market types accord-ing to the way transactions are carried out.

One method of transacting is *competitive trading*, in which many buyers and sellers meet in market places or exchanges. No single seller or buyer has control over the market, and prices are agreed according to rules set by the authority. For example, stocks and shares are sold under the strict supervision of the Stock Exchange. The second method of trading is *bidding*, in which products are sold to the highest bid, or work given to the lowest tender. This is, of course, the method commonly used in construction to find and hire contractors to carry out the work. The third method is *cross trading*, in which buyers negotiate with sellers. This occurs, for example, during the sale and purchase of housing, where negotiations between seller and buyer can take several months, depending on the personalities and negotiating strengths of the two parties involved. The seller may need to dispose of property while the buyer may want the property but lack sufficient funds. When no similar properties are available, agreeing a final price can only be arrived at through negotiation.

Morishima also distinguishes between fixprice and flexprice markets. He notes that the conventional analysis of supply and demand relies on changing prices. Price signals indicate if goods are in surplus or shortage, and prices show the relative values of different products or services. However, this approach implies that prices are changing, continually rising and falling in response to market pressures. This indeed occurs in some markets: share prices and foreign currencies, for example, are in a continual state of flux, changing from minute to minute throughout the trading day. On the other hand, Morishima points out that, in manufacturing, firms tend to keep their prices as steady as they can, but vary their production when markets do not absorb their output sufficiently. Only as a last resort do firms reduce their prices, as this would tend to upset existing customers. Rather than raise their prices with the risk of reversing the increase in the near future, firms prefer to increase output or create waiting lists. Products tend to be sold in fixprice markets, where prices are set by manufacturers, although retailers can vary prices if they wish.

Morishima's framework of types of transaction can be applied to the construction industry. While aggregates are supplied to the market by numerous firms, because the product is non-differentiated and can be used on many sites, there is a system of competitive trading in the basic raw materials for construction as buyers scramble for supplies and sellers compete for sales. As noted earlier, the tendering process is a form of auction, while negotiation or cross trading is more characteristic of the property market. Negotiation may also occur when specific prefabricated components are required. Examples of both flexprice and fixprice markets exist in construction: the labour market tends to be flexprice in nature, as wage rates change frequently to reflect local labour market conditions, while manufactured components tend to be fixprice markets.

Transaction costs

In the property market, the price of a building is only part of the cost. Finance or mortgage payments are affected by the rate of interest. Legal fees, stamp duty and other expenses need to be taken into account. All these costs are associated with the transaction and are in addition to the selling price. Moreover, there are costs involved in finding a suitable property to buy. Searching the market is time-consuming and many offers and negotiations may be unfruitful before a successful outcome is reached.

Williamson (1975) devised a systematic general approach to analysing the costs associated with transactions. These costs arise whenever two firms or individuals interact in a market. For example, firms attempt to make rational decisions concerning price, quality and delivery, but because of the complexities and uncertainties involved it is not possible to be perfectly rational. At some point a decision needs to be made even though some information may be lacking. This aspect of decision-making is called by Williamson *bounded rationality*. In any case, information on technical data, price and availability of alternatives may be deliberately withheld by one firm when dealing with another. This leads to firms transacting at a higher price than would otherwise have occurred. Williamson refers to this as *information impactedness*.

Another source of transaction cost is *opportunism*. This arises as a result of one firm taking advantage over another because of the difficulty of comparing offers. In construction, opportunism arises when firms tender for work knowing that the tender price is less than the eventual price they expect to earn from a job. The tender price enables the firm to obtain the work. Thereafter, it can take advantage of unexpected difficulties that arise, or variations in the contract during the construction phase. Two other related transactions costs associated with markets are *cheating* and *surveillance and control costs*. Clearly, it is always possible for one firm to mislead another by providing a lower-quality product or service than was agreed. Similarly, firms may find excuses to delay payment for deliveries or services rendered. To minimise the cost of cheating firms need to install systems of monitoring and control, which are themselves an additional cost of any transaction.

Transaction costs show that prices only form a proportion of the true costs incurred in market exchange. This is especially important in construction, where the process of production is divided between several firms on site and various professional practices. As a result, transaction costs tend to be high.

9 Basic Accounting Concepts

Introduction

Accounting techniques are equally applicable to firms and to projects within firms. For example, cash flows concern flows of money into and out of company accounts as well as the costs and revenues associated with individual jobs.

Accountants seek to answer a variety of questions, such as: what are the profits that can be distributed to the partners or shareholders of a firm? What are the tax liabilities of the firm and how can they be minimised? What is the value of the assets of a company at the year end? These questions, or more especially the answers, are necessary for a bank or other financial institution to establish the creditworthiness of a particular company or borrower. So far as tax liability is concerned, this has to be agreed between the practice through an accountant and the tax authorities. In order to do that, accepted accounting conventions are used, interpreted and agreed. From an economist's point of view these conventions often appear to be arbitrary.

Accountants have, in effect, set up over the past 500 years a system of recording transactions. This accounting system is used by all businesses and is therefore relatively practical, consistent and useful for the purpose of making comparisons over time and between firms. Certainly, any system is preferable to no system at all. There are four main sets of accounts which summarise different aspects of the accounting process, namely the trial balance; the profit and loss account; the balance sheet; and the sources and application of funds statement.

Apart from reporting on the past performance of companies, accountants also consider budgets, which are concerned with allocating financial resources to meet future needs. In this chapter, cash flow tables describe the timing of inflows and outflows of money and are used to plan and anticipate the timing and size of a company's future monetary requirements. In Chapter 12, discounting techniques are introduced into cash flows to establish the viability of particular projects. The first part of this chapter establishes the difference between credits and debits, creditors and debtors. The second part describes a set of company accounts and the third part discusses some of the ratios used to analyse and monitor the performance of companies.

Accounts may be divided up into different categories, so that detailed information about every aspect of a firm may be kept separately. *Real accounts* relate to assets and liabilities. Assets are those items owned by the firm, including cash, machinery and possibly the premises, as well as money owed to the firm by those indebted to it, namely its debtors. Liabilities, on the other hand, include money owed by the business to its creditors or outside firms, who might have supplied materials or labour on credit. Money owed by the firm to its owners is also a liability on the firm. This money is the firm's capital.

Nominal accounts record wages and other expenses as well as revenues. All real and nominal accounts together make up the full set of accounts of any company or organisation.

Book-keeping

Double entry book-keeping forms the basis of the logic of accountancy. Each category of activity is monitored separately and given its own account. Each account contains entries, which are either debits or credits. A debit entry in an asset account is an increase in the value of the account. For instance, if £5000 is spent on an item of equipment, the value of machinery purchased is an increase in the value of that asset and is a debit in the machinery account, as illustrated in Table 9.1.

A credit entry represents a decrease in an account. Thus the money to pay for the machinery is entered as a credit in the bank account, as cash from the bank is used to pay for the transaction. (This 'bank account' is part of the firm's own accounting system and is distinct from the firm's bank account in its bank.) A debit in one account is therefore also recorded as a credit in another, hence the term 'double entry'. Every time an entry is made in one account, a second entry must be made in another. In this way it is always possible to see from whence a debit (increase) comes and to where a credit (decrease) goes. The £5000 therefore also appears in the bank account, as shown in Table 9.2.

In order to remember the principle of debits and credits, bear in mind that bank statements are written from the point of view of banks. Thus, when a

Table 9.1   A debit entry

	Machinery Account					
	Debits				Credits	
Date	Name	£		Date	Name	£
Jan 1	Bank	5 000				

Table 9.2   A credit entry

	Bank Account				
	Debits			Credits	
Date	Name	£	Date	Name	£
			Jan 1	Machinery	5 000

personal bank account is in credit, the bank owes the customer. Most debits are expenditures in return for assets, while most credits are revenues in return for accepting liabilities. However, not all debits and credits represent cash payments and receipts. Profits are not necessarily the same as the cash difference between money flowing in and out of the firm. Some entries in the accounts are purely notional. For instance, depreciation is calculated but no money changes hands. Similarly, provision for taxation is the amount set aside for future payments of tax and must be deducted from gross profits, even though no payments have yet been made. Valuations of property are based on what a building would be worth if it were sold, even when there is no intention to sell. It is therefore vital to understand that, to accountants, profits are calculated after all costs as well as provisions for tax and depreciation have been deducted from total revenues. Figure 9.1 shows a model of the terms used to describe flows into and out of a firm.

The trial balance

Having discussed the flows of cash into and out of a firm, a set of accounts summarises the transactions which take place in a given period. The first stage

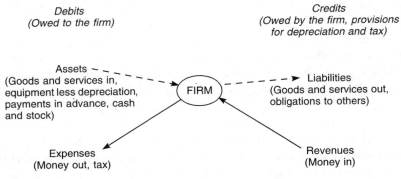

Figure 9.1   Debts and credits

in producing a set of accounts is the trial balance. The trial balance shown in Table 9.3 is used to check that transactions have been properly recorded arithmetically. Making use of double entry book-keeping, the trial balance shows the total of all credits equal to the total of all debits, although the individual entries appear in different accounts. The trial balance simply summarises each ledger or account and sets out the total.

The profit and loss account

A second set of accounts is the profit and loss account, used by accountants to assess the financial results of the work carried out over a period of time, usually one year (see Table 9.4). This provides a figure for the net profit, or taxable profit. Net profit is calculated by deducting all other costs from the gross profit figure. In a limited company, corporation tax is levied on net profits. In a partnership, the net profit is taxed as income tax.

Accountants divide costs into direct costs such as materials, labour and energy used in the production of each item or on each construction project, and indirect costs, which are needed to support the firm as a whole, such as head office costs, including rent, secretarial salaries and other expenses, including insurance and interest payments. Each time a sale is completed, the excess of revenue over direct costs is a contribution, which must be used to pay indirect costs. Only after indirect costs have been met are profits generated. If the practice has had a successful year in business, the result will be a profit, as the revenues from all sources will be greater than the total costs over the period of time concerned.

Any profits made in a given year will be added to the assets of the company and will appear in the balance sheet. Any losses will also appear in the balance sheet and would represent a dwindling of assets. Depreciation of assets purchased in the course of the year will appear in the profit and loss account as a cost or debit item.

Annual accounts are in a sense artificial divisions of a company's ongoing activities. Inevitably there is an overlap between the previous year and the succeeding year. Amounts remaining from the year before are known as accruals. Prepayments are payments set aside for the following year. Adjustments to the accounts for accruals and prepayments have to be made in order to calculate profits. Accruals may arise for fees paid to the firm in the course of a year for work completed in the previous year. Similarly, prepayments, such as rent paid in advance, overlap the following year, and this proportion should therefore also be deducted from these payments before calculating the operating costs for a particular period.

Table 9.3   Trial balance

Trial balance as at 31 March 199–	Debit balances (£)	Credit balances (£)
Capital Accounts		
1st partner		5 000
2nd partner		3 000
3rd partner		3 000
Drawings Accounts		
1st partner	45 000	
2nd partner	23 000	
3rd partner	22 000	
Office furniture and equipment at cost	23 000	
Depreciation of office furniture and equipment – P & L a/c	1700	
Provision for depreciation of office furniture and equipment		16 000
Work in Progress at start of year	75 000	
Debtors and payments in advance	90 000	
Cash at bank	40 000	
Cash in hand	1 000	
Bank overdraft		6 000
Creditors and accrued expenses		72 000
Provision for taxation		2 750
Fees received		600 000
Salaries–architects	160 000	
Temporary architect staff	60 000	
Administration staff	27 000	
Drawing office and modelling materials	12 000	
Publications, subscriptions and exhibitions	5 000	
Travelling and entertainment	7 200	
Carriage outwards	1 200	
Motor vehicle expenses	5 300	
Rent and rates	32 000	
Lighting and heating	6 200	
Repairs and renewals	3 100	
Cleaning	2 100	
Insurance	21 000	
Telephone	9 000	
Stationery and printing	20 000	
Postage	2 500	
General expenses	3 250	
Bank interest and charges	3 600	
Accountancy	6 000	
Legal and professional	600	
	707 750	707 750

Closing Work in Progress at 31 March 199– amounted to £75 000.

Table 9.4   Profit and loss account

Profit and loss account for the year ended 31 March 199–			
	£		£
Architects' salaries	160 000	Fees received	600 000
Temporary architect staff	60 000		
Materials	12 000		
Rent and rates	32 000		
Salaries (Secretarial, etc.)	27 000		
Travelling expenses	7 200		
Carriage outwards	1 200		
Motor vehicle expenses	5 300		
Telephone	9 000		
Lighting and heating	6 200		
Stationery and printing	20 000		
Postage	2 500		
Exhibitions, subscriptions	5 000		
Insurance	21 000		
Repairs and renewals	3 100		
Cleaning	2 100		
General expenses	3 250		
Depreciation	1 700		
Bank interest and charges	3 600		
Accountancy	6 000		
Legal and professional	600		
Stationery and printing	20 000		
Net profit to capital a/c	211 250		
	600 000		600 000

The balance sheet

The balance sheet is a summary of the assets and liabilities of a firm on a particular day, usually at the end of a year's trading. All assets and liabilities are summed. If total assets are greater than total liabilities, as shown in Table 9.5, the difference is a net profit. If, on the other hand, total liabilities are greater than total assets then the firm will have incurred a net loss, as in Table 9.6. By definition, a balance sheet always balances.

Assets include plant and equipment. However, in the course of a year, assets depreciate because of wear and tear. Accountants use various techniques to allow for this adjustment in the book value of assets. One such technique is *straight-line depreciation*. The anticipated second-hand or scrap

Table 9.5   Balance sheet showing net profit

Assets	£1000		Liabilities	£700	
			+ profit	300	
		£1000			£1000

Table 9.6   Balance sheet showing net loss

Assets	£1000		Liabilities	£1200	
			− loss	200	
		£1000			£1000

value of the asset after a number of years is deducted from the purchase price. The difference is then divided by the number of years. This figure is then subtracted annually from the remaining value of the asset in order to give a current valuation.

Depreciation must be taken into the profit and loss account to show the decline in value of assets used up in the course of the year. Depreciation must also be deducted from assets in the balance sheet, to record a more realistic valuation than the purchase price paid. For example, assume a car cost £10 000 and at the end of two years is to be sold for an expected selling price of £4000.

$$\text{Annual depreciation} = \frac{£10\,000 - £4000}{2} = \frac{£6000}{2}$$
$$= £3000$$

Deducting an amount for depreciation each year recognises the charge to profits for the use of the asset during the period. It is as if the funds will be used to replace the car at the end of its useful life. Otherwise, the practice will have consumed its own capital, and at the end of the two years the capital will have declined from £10 000 to only £4000.

An alternative method of depreciating an asset is the *revaluation method*. The car in this example is given a value at the end of each year based on its second-hand price at the time, and the annual drop in value is the figure allowed as depreciation. This method takes account of the fact that many items of equipment depreciate rapidly in the first year or two. If the car would only fetch £6000 after one year, then depreciation in the first year is £4000. If, after two years, the car could command £4000 in the second-hand car market, then depreciation in the second year is £2000.

There are other methods of depreciating assets, but most are as arbitrary as the straight-line method. In general, economists prefer the revaluation method, as it relates to the concept of opportunity cost and can be used for decision-making. If an asset is worth £5000, then the opportunity cost of keeping it is the next best alternative that can be obtained if it is sold. The choice is then between the asset and its alternative.

Buildings and sites are fixed assets in the balance sheets of many firms. To property companies, buildings and sites are used as stock to be bought and sold in the course of the year in order to generate profits. Property valuation is therefore important for calculating the assets of a firm, and the buying and selling prices of property companies themselves. Moreover, the starting point for the evaluation of projects is the value given to completed buildings. This figure can then be used to budget construction costs and the value of sites. The issue of property valuation will be dealt with in Chapter 12.

Assuming that the assets are all given their realisable values (that is, those values that would actually be paid if the assets were sold), the capital value represents the value of the business and can be calculated by summing the assets and subtracting the debts owed to banks, staff and other suppliers, who together constitute the creditors of the business. In other words, the amount left after paying everybody else, including tax due, is the worth of the business to its owners.

Spending on expenses is called current expenditure and includes such items as electricity, stationery and rent, which are all used up in the course of providing the service. Current expenditure is treated differently from capital spending on durable assets, such as plant and equipment. Capital expenditure is used to purchase assets which have a value lasting more than one year, and which can therefore often be resold, whereas current expenditure is for items which cannot be resold such as wages and salaries.

Current assets consist of the value of debtors, prepayments and cash. To ensure a firm's liquidity, the total of current assets should be greater than current liabilities, namely the total owed to creditors, the value of accruals and the size of any bank overdraft. Net current assets are assets less creditors and accrued liabilities. In principle this represents a valuation of the assets available in the event of liquidation. However, the valuation of assets given in the balance sheet may not necessarily be realisable. Their actual value would depend on the price at which they were sold. Table 9.7 is an example of a balance sheet.

Profits appear as a liability in the balance sheet because accountants argue that the business owes this sum to the owner of the firm. Architectural and surveying practices, for example, most often take the form of a partnership, which was defined in the Partnership Act (1890) as 'two or more persons

Table 9.7   A balance sheet

Balance sheet as at 31st March 199–			
	£	£	£
Assets			
Fixed assets			
Property, furniture and fittings	23 000		
Less depreciation	16 000		
		7 000	
Current assets			
Work in Progress at 31 Mar 199–	75 000		
Cash	1 000		
Bank balance	40 000		
Debtors	90 000		
		206 000	
			213 000
Financed by			
Liabilities			
Capital accounts			
1st partner	5 000		
2nd partner	3 000		
3rd partner	3 000		
		11 000	
Current accounts			
1st partner (106 250 − 45 000)	61 250		
2nd partner (53 000 − 23 000)	30 000		
3rd partner (52 000 − 22 000)	30 000		
		121 250	
Current liabilities			
Bank overdraft	6 000		
Creditors and accrued expenses	72 000		
Tax	2 750		
		80 750	
			213 000

carrying on business with a view to profit'. In a partnership, these profits may be appropriated or shared between the partners according to their particular partnership agreement. If the distribution of profits is not specified in an agreement, profits are shared equally between the partners. Additional partners alter the partnership agreement and the profit-sharing ratios.

Partnerships are very much private affairs. There is no need to publish accounts or publicly state the aims of the business. It is not a separate legal entity from the partners themselves, each of whom is personally responsible for the firm's debts. This is not necessarily a problem if the partnership does not rely on large credit from suppliers of stock or equipment. However, if a firm has extended lines of credit, limited liability can be used to protect the principals from being sued for financial losses incurred by creditors in the event of non-payment or late payment, unless personal guarantees are required.

A limited company is a separate legal entity from its owners, the shareholders. Having supplied the permanent share capital, shareholders receive ordinary shares. Their investment in the firm will not be repaid, unless the company is wound up and surplus assets distributed. Shareholders can only sell their holdings to other individuals, who then become the new shareholders. Those profits not retained in the business are distributed in the form of dividends to the ordinary shareholders. These dividend payments vary depending on the firm's financial performance and decisions taken by the directors.

Preference shares are shares with a fixed dividend, which must be distributed before ordinary shareholders may receive their dividend payments. Debentures, on the other hand, are usually equivalent to a mortgage, earning interest until the repayment date.

Creditors have access to financial information about limited companies at Companies House. Various acts of Parliament, including the Companies Act (1985), require limited companies to publish annually the audited final accounts, namely the profit and loss account and balance sheet. An alternative way of presenting the balance sheet of a limited company is shown in Table 9.8.

Table 9.8   Balance sheet of limited company

Balance sheet as at 31 March 199–					
Net worth	Liabilities			Assets	
	£	£	*Fixed assets*	£	£
Preference shares	32 250		Office furniture	17 000	
Ordinary shares	100 000		Drawing office equip.	6 000	
		132 250	Less depreciation	16 000	
					7 000
Current liabilities			*Current assets*		
Accounts payable	72 000		Cash	41 000	
Debt	6 000		Debtors	90 000	
Other liabilities	2 750		Work in progress	75 000	
		80 750			206 000
Total		213 000			213 000

The sources and application of funds

The last summary financial statement describes the allocation of cash flows. The sources and application of funds statement shows the profits for the period, how these are used to purchase assets or repay loans, and any resulting increase or decrease in working capital. Table 9.9 is a sources and applications statement based on the data in Table 9.3.

Thus the sources and application of funds statement shows the effect on working capital of management decisions, either clearing the way for further expansion or indicating that measures may be needed to protect the firm's cash flow.

Working capital

A set of accounts may be used to measure the results of the activities of a firm. It is arguable that the purpose of the accounts is also to allow

Table 9.9   Sources and application of funds statement

Statement of sources and application of funds for year ending 31 March 199– Source of funds	£	£	£
Details including:			
1. Profits before tax	211 250		
2. Adjustment for depreciation as this will have been deducted from profits (non-cash items)	1 750		
TOTAL		213 000	
Less application of funds			
Details including:			
1. Repayment of loans			
2. Distributed profits	90 000		
3. Tax paid	2 750		
4. Purchase of fixed assets			
Increase (or decrease)			116 750
Increase (or decrease) in working capital			
Details including:			
1. Change in debtors	40 000		
2. Change in creditors	50 000		
3. Change in bank	26 750		
Increase (or decrease)			116 750

shareholders of companies or partners in professional practices, as well as potential investors, to evaluate the assets of a firm. Directors and partners can see if there has been any change in the value of the firm's assets from one year to the next. In order to extract as much information from accounts as possible, it is necessary to carry out an analysis of the figures. For example, working capital is the excess value of current assets over current liabilities. In Table 9.8, working capital is equal to £206 000 less £80 750, namely £125 250. Working capital is needed to ensure that there is always sufficient money available to meet payments in the short term. A shortage of working capital is often the cause of firms running into cash flow difficulties, forcing them into liquidation even when they are busy, trading profitably and expanding. Indeed, a shortage of working capital is often caused by expanding too rapidly.

When firms grow, they need to take on additional staff, and buy extra materials in advance of payment from clients. If sufficient funds are not available, the company would be unable to pay staff or suppliers in spite of having a large order book. This cash flow problem is known as *overtrading*, and frequently occurs in construction when small firms take on more work than they are capable of financing.

A shortage of working capital also follows when payments to firms are delayed. The firm simply runs out of cash to pay wages and other short-term debts which cannot wait until clients settle their invoices. Indeed, as soon as a firm runs into cash flow difficulties, its clients become increasingly reluctant to pay for unfinished work, further adding to its problems.

Ratio analysis

Ratio analysis is used to study accounts. A ratio is simply one item in a set of accounts divided by another. Any two items can be used to form a ratio. Ratios are used to compare the trends in a company's performance from one year to the next, or to compare different companies, regardless of their size. A very useful source of information on the performance of firms in the construction sector is provided by Inter Company Comparisons Business Ratio Reports. These reports use a variety of accounting ratios, including some of the ratios discussed below, to analyse the performance and growth rates of a sample of firms in the same industrial classification.

The acid test or quick ratio

One such ratio is the ratio of current assets (excluding stocks) to current liabilities, given in Equation (9.1). This important ratio is used to assess whether or not a firm has sufficient working capital to continue trading. It is a measure of the financial health of a firm and is called the acid test:

$$\text{The acid test} = \frac{\text{Total current assets}}{\text{Total current liabilities}} \times \frac{100}{1} \qquad (9.1)$$

If the acid test is greater than 100, then sufficient sums are available or expected shortly to meet short-term requirements. In other words, the firm is solvent; it can honour its obligations. If the ratio is less than 100, then cash flow difficulties can be expected.

The ratio may also be too high if the amount in debtors are allowed to increase beyond a reasonable limit. A satisfactory ratio might lie between, say, 150 and 200 per cent. The higher the ratio, the more working capital is required to finance it.

Taking the data from the balance sheet in Table 9.8, and substituting into Equation (9.1):

$$\text{The acid test} = \frac{206\,000}{80\,750} \times 100$$
$$= 255.1\%$$

The acid test in this instance does appear to be high. This suggests that reducing the amount owed by debtors would reduce the amount of working capital needed to finance the firm. With less working capital required, outstanding loans and overdrafts could be reduced and this could lower the firm's interest payments on its own borrowing. If there were no outstanding loans, then receipts could be invested or distributed to shareholders.

The ability to meet immediate financial obligations depends on the cash position of a firm. This is given by the following ratio:

$$\text{Cash position} = \frac{\text{Total cash}}{\text{Total current liabilities}} \times \frac{100}{1} \qquad (9.2)$$

Clearly, too little cash will mean the firm cannot meet its obligations in full, while too much cash will mean the firm is forgoing interest or dividends that could be earned with the excess cash. In the above example, the cash position is therefore:

$$\text{Cash position} = \frac{41\,000}{80\,750} \times 100$$
$$= 50.77\%$$

In this case, the firm has just over half the cash needed to meet all its current liabilities and unless it is paid soon, the firm will need to sell some of its assets, forestall creditors or borrow money. The absolute sums of money involved are not as significant as the relative proportions of various assets and liabilities.

Performance ratios

The cash position of a firm is affected by the credit terms it offers its customers and the period of time it takes to pay its suppliers. In other words, a company's operating effectiveness and financial control may be indicated by examining its performance ratios. An extremely important performance ratio for professional practices, main contractors and subcontractors may be used to estimate the average number of days taken by clients (debtors) to settle their accounts. This is calculated as follows:

$$\text{Number of days} = \frac{\text{Debtors}}{\text{Credit sales for the year}} \times \frac{365}{1} \qquad (9.3)$$

From the profit and loss account and balance sheet, this ratio is:

$$\text{Number of days} = \frac{90\,000}{600\,000} \times \frac{365}{1}$$
$$= 54.75.$$

It therefore takes on average 54.75 days for this particular firm to receive payment, assuming the £90 000 of debtors at the end of the year is representative of the average level of debtors throughout the year.

A very slow rate would mean that the firm not only financed its clients with free loans, but was also paying interest on the outstanding money the firm must have spent to finance its work on projects on behalf of its clients. At the very least, the outstanding amount would not be earning interest for the benefit of the firm.

Slow payment by clients is a major problem facing small firms and professional practices, who have little bargaining power, and who depend on good relations with their customers for future work. One of the major issues discussed in the Latham Report (Latham, 1994) is the problem of 'pay when paid' clauses in construction contracts between main and subcontractors. These clauses enable the main contractor to delay payment until paid by the employer or developer. Nevertheless, different firms adopt a variety of procedures to deal with the problem of late payment. These range from penalty clauses to charging interest on overdue amounts. However, these sanctions are often difficult, impractical and costly to impose.

One system adopted by many firms at least enables the firms themselves to monitor their cash flow situation. Towards the end of the month following that in which an *invoice* was issued, a *statement* is sent to the client. In the third week of the following month a *first reminder* is sent. If settlement has not been received by the third week of the third month, it may be necessary to phone the client to enquire about the reason for the delay or to send a letter stating that there is no alternative but to put the matter in the hands of a third party, usually a lawyer. Of course, this procedure should always be handled with discretion and flexibility.

A second performance ratio is the working capital turnover ratio:

$$\text{Working capital ratio} = \frac{\text{Turnover}}{\text{Working capital}} \times \frac{365}{1} \qquad (9.4)$$

From the final year-end set of accounts, the working capital turnover ratio is:

$$\text{Working capital ratio} = \frac{600\,000}{125\,250} \times \frac{365}{1}$$
$$= 1749$$

This ratio shows how many days' supply of working capital the company has at its disposal. It shows if more working capital is required to allow for late payment, expansion or contingencies.

Similarly, the asset turnover ratio is given by the formula:

$$\text{Asset turnover ratio} = \frac{\text{Turnover}}{\text{Capital employed}} \qquad (9.5)$$
$$= \frac{600\,000}{132\,250}$$
$$= 4.54$$

This ratio relates the capital invested in a firm to its turnover or volume of sales. If this ratio is declining, then pound for pound the firm is no longer generating the same amount of business out of the assets it has at its disposal. This would give an early indication that measures were needed to reverse the trend. If, for example, more capital were invested in the firm, one would expect to see increased sales. If, however, the sales have not risen in proportion to the injection of capital, then decreasing returns may have set in. This, in turn, could imply that the firm is approaching its optimum size and further expansion is not desirable. Progress and expansion plans would have to be monitored carefully.

Profitability may also be measured. The gross profit margin provides a useful comparison between companies and from one year to the next. It is given by the formula:

$$\text{Gross profit margin} = \frac{\text{Gross Profit}}{\text{Sales}} \times 100 \tag{9.6}$$

The gross profit should be high enough to cover all overhead expenses. If the margin is increasing in the above ratio, this will show a healthy improvement over previous years. Otherwise, future difficulties may possibly be anticipated.

An alternative example measuring the profit as a percentage of sales is the net profit margin. This gives a more precise picture of profitability after all expenses have been met:

$$\begin{aligned}
\text{Net profit margin} &= \frac{\text{Net profit after taxes}}{\text{Sales}} \times 100 \\
&= \frac{211\,250 - 2750}{600\,000} \times 100 \\
&= \frac{208\,500}{600\,000} \times 100 \\
&= 34.75\%.
\end{aligned} \tag{9.7}$$

Another useful ratio concerned with profitability relates fee income to direct labour costs. Direct labour costs should only be a relatively small proportion of the fees charged, so that the fee income is sufficient to cover all other office costs and leave a profit. If the multiple begins to decline, fee charges may have to be increased:

$$\begin{aligned}
\text{Fee multiple} &= \frac{\text{Fee income}}{\text{Direct labour costs}} \\
&= \frac{600\,000}{220\,000} \\
&= 2.73
\end{aligned} \tag{9.8}$$

Gearing ratios

Gearing ratios are also useful indicators of a firm's exposure to risk. They relate a firm's assets to its debts. Thus, if a firm has a high gearing ratio it runs a high risk of being unable to service its debts, either to pay the interest or eventually to repay the debts themselves. Because relatively little capital is required to set up an architectural or surveying practice, or even a building contractor in business, long-term debts are usually minimal. Gearing ratios will therefore usually be more relevant to establishing a client's creditworthiness than to an architect's or surveyor's own firm. Nevertheless, the total gearing ratio is:

$$\text{Total gearing ratio} = \frac{\text{Long-term debt} + \text{current liabilities}}{\text{Total assets}} \times 100 \qquad (9.9)$$

An alternative to the total gearing ratio is the long-term gearing ratio which is:

$$\text{Long term gearing} = \frac{\text{Long-term debt}}{\text{Total assets} - \text{current liabilities}} \qquad (9.10)$$

Whichever gearing ratio is used, the higher the percentage, the greater the risk.

Cash flow tables

Cash flow is the difference between money entering and money leaving a firm. A positive cash flow increases a firm's liquidity, namely the amount of cash at its disposal. Negative cash flows occur when the outflow of cash is greater than the inflow. Cash flows measure and highlight when an expected negative cash flow is likely to be at its greatest during a project. This can be used to anticipate and plan cash requirements to avoid difficulties.

Cash flow difficulties can be overcome by making provisions – by, for example, seeking overdraft facilities at a bank well in advance. A cash flow table might well be used to accompany a request for overdraft facilities at a bank as it shows the timing and size of the loan required and its expected repayment schedule. The following cash flow relates to a firm, but similar methods can be applied to individual projects. The figures in Table 9.10 might have represented the six months following the above set of accounts.

From Table 9.10, it can be seen that the balance at the start is £41 000. This figure comes from the balance sheet. To the £41 000 is added the income for the first month. Next, the expenses for that month are deducted, to leave the expected cash position at the end of the month. This figure is then carried forward to the following month to repeat the same accounting procedure. At the end of the second month, a cash flow deficit can be observed but this is not a problem as the table shows a positive cash flow in the following periods. If negative cash flows are anticipated, then cash flow crises can be avoided.

Table 9.10   Cash flow statement

Period	1 £	2 £	3 £	4 £	5 £	6 £
Income						
Fee income	40 000	20 000	150 000	10 000	60 000	35000
Other income						
Total income	**40 000**	**20 000**	**150 000**	**10 000**	**60 000**	**35 000**
Expenses						
Professional salaries	25 000	25 000	20 000	16 000	16 000	16 000
Clerical salaries	2 250	2 250	2 250	2 250	2 250	2 250
Materials		2 000		2 000		2 000
Stationery, printing	1 800	1 800	1 800	1 800	1 800	1 800
Fixtures and fittings			4 500			
Equipment				10 000		
Subscriptions, exhibitions				400		3 000
Travel	500	500	500	500	500	500
Entertainment	500	500	500	1 000	500	500
Motor expenses	600	600	600	600	600	600
Rent, rates	20 000					
Insurance		12 000				
Services	1 000		1 000		1 000	
Telephone		2500			2500	
Postage	200	200	200	200	200	200
Repairs and renewals			1 500			1 500
Cleaning and sundries	400	400	400	400	400	400
Drawings	4 000	4 000	10 000	10 000	10 000	10 000
National Insurance	2 500	2 500	2 200	2 000	2 000	2 000
Bank charges		2 000				
Accountant				7 500		
Total expenses	**58 750**	**56 250**	**45 450**	**54 650**	**37 750**	**40 750**
Balance at start	**41 000**	**22 250**	**(14 000)**	**90 550**	**45 900**	**68 150**
Add Income	40 000	20 000	150 000	10 000	60 000	35000
	81 000	42 250	136 000	100 550	105 900	103 150
Less Expenses	58 750	56 250	45 450	54 650	37 750	40 750
Balance at end	**22 250**	**(14 000)**	**90 550**	**45 900**	**68 150**	**62 400**

10 Planning and Control of Firms and Projects

Introduction

Firms and organisations in construction are project-orientated. Unlike firms in manufacturing, construction firms rarely do the same thing in the same place from one year to the next. Firms undertake projects and allocate staff to carry out the work. Within each firm, several projects must be co-ordinated in order to make the best use of staff and resources, and to provide continuity of work and revenues. The selection of projects according to size and type of work must also be consistent with the general aims and targets of the firm. Otherwise, the firm can only respond in a haphazard way to opportunities as they become available, rather than going out to seek work of a particular type and quality.

The work undertaken by firms in construction cannot be accurately planned, as it depends on circumstances outside the control of the individual firm. Nevertheless, the marketing of a firm depends on the perception of its competence to undertake certain kinds of work, and its role within certain types of contractual arrangements. It is therefore up to the firm to decide if it wishes to be considered for work as a specialist contractor, subcontractor or main contractor, whether in the public sector or the private sector, in which region or area, and on which size of project.

All these decisions concerning the types of project to be undertaken need to be made in the light of a business plan in order to be consistent with the firm's resources, skills, contacts and experience, as well as its plans for the future. For this reason, this chapter deals with both the overall business planning process and the management of individual projects which form the building blocks of the work of contractors and professional practices in the construction industry.

Business objectives

It is necessary to state the aims of an organisation before setting up its system of management. Apart from making profits, other aims include promoting the survival of the firm, giving employment to people, and producing a high-quality product or service. A non-profit-making organisation may have other objectives, but it will still hope to achieve these efficiently and provide value for money.

Goals may be seen as the longer-term general aims of a firm, sometimes called the firm's mission statement. Mission statements often refer to the type and quality of a firm's output, the firm's role in society, and its attitude to its workforce. Objectives are shorter-term goals, such as increasing the number of employees, size of turnover or share of market. Objectives can be measured so that targets can be agreed and performance monitored. Strategies are the means of achieving objectives; for example, diversifying the type of service offered, or expanding the size of the firm. Tactics relate to the day-to-day activities required to support the strategies; for example, deciding on tendering policies and the percentage markup on costs, upgrading computers in the firm, or improving staff training.

Within any commercial organisation there is a need to plan, and planning relates to the level of skills, resources and time required. Wasted time is time lost for ever. Working hard the next day to make up for lost production only means the extra effort could have been used in addition to the lost output.

The business plan

A business plan provides a context in which to understand and co-ordinate the activities of the firm, its targets, aims and objectives. Business plans have practical uses. For example, they may be prepared for banks when external finance is required to set up or expand a firm. A business plan should include the following:

1. There should be a clear statement defining the product or service, including the types and number of building projects, which the firm would hope to carry out in the period of the plan.
2. The targeted market should be defined, as the marketing effort of the firm depends on the kinds of people, firms and organisations to approach. For example, while the public sector must advertise all projects and offer them in open competition, there is no legal obligation on private-sector clients to do so.

3. Prices, fees or rates of charges should also be estimated, as well as the expected workload, to provide a figure for future income. On the basis of these estimates of workload it is possible to calculate the firm's resource requirements in terms of labour, materials and equipment as well as finance.

These figures can then be fed into a cash flow table such as that illustrated in Figure 9.10 on page 194. Cash flow tables can be used to set financial targets from month to month and to monitor progress, as well as revising targets during the course of a year if necessary. A business plan is not a rigid document. It should take account of changing circumstances. The business plan is the link between current performance and future investment decisions.

Corporate strategy

A corporate strategy is a long-term business plan. The basic function of a corporate strategy is to determine the size and scope of activities well into the future. Setting long-term objectives determines the firm's need for investment and the profits needed to finance that investment.

Every aspect of a firm's strengths, weaknesses and opportunities – in both marketing and in technological terms, and threats to the firm, should be considered. This is often referred to as a SWOT analysis (strengths, weaknesses, opportunities and threats) and includes types of project to be undertaken, the aptitudes of staff, and access to funding. External aspects, such as the competition facing a firm, the economic conditions in proposed markets and the general economic climate should also be part of the analysis. Wherever possible statements should be quantifiable, to enable the plan to be monitored. Quantifiable targets, where appropriate, should be set after negotiation and with the agreement of the employees concerned. As with any plan corporate strategies should include completion dates for tasks or projects.

The first step in planning a corporate strategy is to assess the current position of a firm in order to quantify the resources the firm possesses and how it currently operates. The starting point for many corporate strategies includes the value of the assets, liabilities, turnover and profits. A number of issues may also be raised at the beginning. What, for instance, is the firm good at? What could it do, that it does not do at present? What could it improve? The answers to these questions form the basis of the corporate strategy.

As many people in the firm as possible should be involved in the discussions on corporate strategy. Topics raised with staff might include some of the following questions. Is there sufficient communication with employees? Are individual members of staff challenged in their work, and in what ways might they wish to develop? How might the firm help its employees to be more effective in their work?

The second step is to consider the targets to achieve in, say, five years' time, and then to work backwards, stating the figures for the fourth year that would enable the fifth year's objectives to be reached. If one of the objectives was to double the current year's turnover in five years, then one would need to consider the implications from the point of view of the turnover targets in four years' time, three years' time and so on. How much, therefore, should the turnover be in one year's time, if the second year's target is to be fulfilled? Consequences of such planning lead one to take other factors into account, such as the amount of investment required and the profits needed to finance that investment.

The corporate strategy pulls all aspects of a firm together to form a comprehensive plan. There are two features worth noting of such a plan. The first is that it is flexible in that it must be adapted whenever circumstances demand a modification. Second, it has a shifting time horizon. Though the plan states objectives for the years to come, corporate planning is a continuous process of reviewing a company's progress. A year after completing the first year of a five-year strategy, a new five-year strategy is drawn up, asking the same types of question, modifying the first plan and extending it for another year to make a new five-year horizon.

One weakness of this approach to strategic planning is that quantifying work and setting measurable targets has limitations. Measurements often ignore the issues of quality and complexity. Moreover, the productivity of individuals within a firm is determined by factors beyond their own control, such as instructions given by superiors, or unexpected difficulties on site.

Nevertheless, discussion about corporate strategy will tend to improve relations between employees, understanding of the firm and motivation of individuals. It will give employees a better understanding of their individual roles, their contribution to the overall effort and an idea of what is expected of them. After all, the success of a firm depends on and represents the work and effort of every member of staff.

The financial management of projects

Having discussed organisational matters in a broad context, we now turn our attention towards various specific techniques used to plan, organise, manage and control work on particular projects. A cash flow table of expenditure during a project can be used to anticipate construction costs and when payments are likely to arise. The general pattern of payments from one project to the next is remarkably similar, although the variations from project to project can be great, and the timing and size of payments in one job cannot be predicted on the basis of other projects.

Generally, the cash flow starts slowly, with small payments being made for professional work carried out in the early stages. The size of payments increases when work commences on site and continues to increase as more and more labour, subcontractors, materials and hired plant are brought into the programme. Towards the end of the construction phase most of the structural work is complete, and the amount of heavy equipment needed and the quantity of materials arriving on site as well as the size of the work force declines. The number and size of payments towards the finish also tend to decline. The pattern of these payments over time is similar to a normal distribution as shown in Figure 10.1.

During the period of construction, three sets of costs need to be closely monitored:

1. The actual cost of work performed.
2. The budgeted cost of work scheduled.
3. The budgeted cost of work performed.

The actual cost of work performed must be known in order to measure progress and ensure a satisfactory payment system. A budgeted cost of work scheduled is used to monitor progress to anticipate possible difficulties

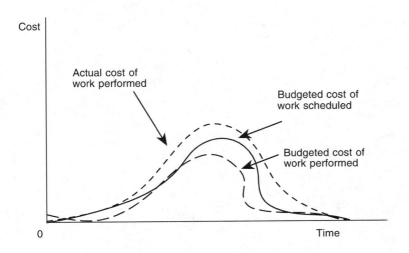

Figure 10.1   The distribution of costs during the construction phase

in, for example, co-ordinating the work of various subcontractors. Finally, the budgeted cost of work performed is used to compare with the budgeted cost of the work scheduled in order to assess the schedule variance. The schedule variance indicates the difference between the time allowed and that actually used. The cost variance is the difference between the budgeted cost and the actual cost of the work performed. By the end of the project these differences become the project's overrun.

As work progresses on site, costs accumulate. At the beginning costs are minimal but as structural work gets under way total costs to date start to rise rapidly. Later, as building work nears completion and finishing trades gradually leave the site, the rate of increase in total costs slows down the pattern of costs accumulating over time forms an S-shaped curve as shown in Figure 10.2.

Fast track

Fast track means rapid construction by overlapping design and construction phases. This implies doing away with the assumption that one phase must be completed before the next phase can commence. Drawing may begin even before planning permission has been granted. Naturally, the risk of aborting work is greater, using fast track techniques. To minimise such risks, professional

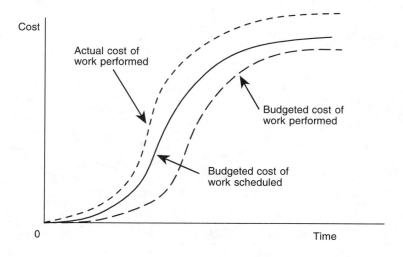

Figure 10.2   The S-curve: the cumulative distribution of costs during the construction phase.

advice at every stage demands team effort and co-operation between the different advisers, management, contractors and labour. Construction phases can also be overlapped so that the critical path may be shortened. For example, fire-proofing does not need to be completed before cladding can commence on another part of the building.

Carpenter (1982) points out several advantages of reducing the building period, including savings on time-related costs such as hiring plant, reductions in interest payments because funds are borrowed for a shorter period, and a reduction in the effects of inflation on raising costs. He notes other less obvious benefits of fast track, such as avoiding changes in clients' requirements, and changes in the administrative and political environment. Moreover, early completion, according to Carpenter, enables the client to gain a competitive advantage over other firms by being in business before them.

The use of computers and information technology

The impact of computers has been mainly in terms of cost reduction and increased output at the expense of longer-term job security. The major cost reduction is the result of improved management techniques and more accurate information. It is true that fewer employees are needed now than were previously to carry out the same quantity of work, especially much of the routine or repetitive clerical and draughting work. Even standard letters and reports can be personalised, thus reducing secretarial overheads. However, the second impact of the new technology has been to increase the number of drawings a practice can produce, and the greater flexibility of a computerised design system has not so much reduced the work of the office, but rather has enhanced the quality and reliability of the service offered. Use of new technology can also save time.

Information technology has altered the way management functions throughout industry. There has been a shift away from simply directing a large number of routine tasks carried out by a large number of unskilled labourers, towards far more complex and informed decision-making related to resources and methods of production. This trend in industry may be reflected in the architectural and surveying professions as well as in contracting firms themselves, as more powerful computers and software force managers of the building process to adopt new methods. There are changes in the way offices are managed, changes in the draughting work of the architect, and broader approaches to tackling projects and feasibility studies.

One of the biggest problems associated with the introduction of new technology is the relationship between staff and computers. Computers and computer software often involve lengthy training. There is a widespread fear of new technology, especially among older staff, together with a perceived threat to jobs, especially during a period of recession in the industry. Computers may be used not only to assist employees in their present roles but also to change the way people work. The questions arise: to what extent should people change their methods of working to accommodate the computer? Or should computers be adapted to meet the needs of employees? Using computers permits more design and evaluation to be done simultaneously. In principle, this could avoid costly problems on site. In practice, these benefits have not always been forthcoming. Although design and build increases the potential for co-operation, traditional demarcation lines between professionals often mean that possible savings of time, cost and effort resulting from the use of computers are not often realised. In any case, the gains currently attributable to computing techniques alone should not be exaggerated. They may only have a marginal affect on construction costs compared to other technical and organisational changes that could be introduced. This is not to say, however, that in future the gains could not be significant.

Computers have become increasingly *user friendly*, which means that their use has been simplified to such an extent that little, if any, technical knowledge is required to operate them. Computers have found their way into professional offices in four main areas: computer-aided design (CAD), management of the office, costing of projects, and project management.

Computer-aided architectural design/draughting

Computer-aided design may be used to draw, test and store ideas quickly and efficiently. Drawings can be duplicated, scaled up or down as required, and altered for different members of the construction team. Engineers require modified drawings with the same dimensions but different details. CAD can ensure that all drawings are co-ordinated and consistent. Using computer drawings enables everyone to work from the same basic document, which can be altered quickly and cheaply. Far more drawings can be generated and more options considered in a given period than by the use of conventional techniques. Computers reduce drastically the frequency of errors during the duplicating process. Using computer graphics not only improves the service offered to clients, it is also a useful selling point, creating, often with some justification, the image of an up-to-date and efficiently-run office.

There are three basic types of systems; namely, *two-dimensional*, which present plans and elevations; *three-dimensional*, which provide perspective views; and a combination of the *two and three-dimensional systems*. The last translates two-dimensional drawings into three dimensional geometric per-spectives, and vice versa.

Many CAD systems also measure, count and cost design elements as the design develops. The costings are taken from cost databases, the electronic equivalent of, for example, Spons or Laxtons. This integrated approach to design means that designers are now in a better position to design to budget, and quantity surveyors are able to monitor progress and update bills of quantities as soon as changes are made. Of course, much work still needs to be done in developing the software and new working practices. Nevertheless, using conventional methods, designs could only be submitted to quantity surveyors for costing after the design proposal was well advanced and expensive to alter.

Computerised costing of projects

Costing exercises have to be carried out not only during the design stage but also before and after this. Computer programs are widely used to cost and estimate proposals. They allow alterations to be incorporated easily and quickly whenever a change occurs, either in a supplier's price, the quantity required, or the specification of an element. This saves time and reduces costs as well as the stress, caused when notification of changes arrives at the last minute. Again, computers reduce the risk of error when calculating the extensions of quantities and unit prices in, for example, the preparation of priced bills of quantities.

Trade or work section bills of quantities, in the form used for tendering and on site, can also be converted very easily to elemental bills of quantities, which is of use to the designer allocating costs to various parts of a building according to the priorities of the design. Because the information is now readily available, it is up to architects, quantity surveyors and engineers to make greater use of it. The cost implications of design ideas can be assessed instantly, giving the architect greater confidence in meeting the client's brief within budget. The list of elements in an elemental bill of quantities is given in Chapter 13 as part of the form of a feasibility study.

As well as providing bills of quantities, computer software is used in carrying out feasibility studies to produce discounted costings to test different designs for economic viability and to compare alternatives. The software used is generally based on a spreadsheet. Computer spreadsheets are an electronic version of graph paper. In each square or cell, information can be stored and used in calculating total costs, by specifying which cells should be combined to produce the desired results. Fortunately, several off-the-peg programs have been devised for use in the construction industry. New approaches are also being developed in this field as improved versions and new ways of working are made available by the adoption of computerised information. Different options can be fed into spreadsheets of data to produce the cost implications almost instantly. Computers are used in this way to appraise development proposals and calculate the return on assets employed.

Computerised project management

Effective site management can reduce construction time and building costs more significantly than almost any other single factor, because of the improved co-ordination of labour, machinery and materials, as well as cash flow. Project managers must be able to plan the construction, phase the work and monitor progress and costs. Computers can be used to draw critical paths through networks. The result can then be used to schedule and budget the work. As the data is input only once, the whole programme of events and costs will be consistent, although its accuracy will usually depend on factors quite separate from the computer technology.

Computer-generated S-curves enable managers to compare variances as often as necessary. However, an S-curve only represents one possible scenario. Again, the effect of delays on costs at various stages can be seen in advance. With computers, it is possible to examine the consequences of possible difficulties and calculate the risk of occurrence as a percentage of the most likely outcome.

Using computers can alter the whole management approach, because of the ease of calculating and recalculating each time new information becomes available, or each time decisions need to be taken. The need to rely on intuition and experience has been complemented by a range of mathematical management tools which can evaluate the implications of changing circumstances to help managers make better informed decisions. Software can generate hundreds of *what if* scenarios to allow managers to anticipate the consequences of changes in profit margins, delays in payments and modifications to other factors, which arise in the course of a project (Khosrow-shahi, 1991).

Similarly, computers may be used to assess risk by simulating different *what if* situations, including the financial implications of different rates of inflation and delays caused by adverse weather conditions. In fact, there may be thousands of variables involved in a building project. Consequently, the volume of calculations very quickly becomes unmanageable without the power and speed of a computer. Maximum, minimum and mode values may be fed into the computer to assess the range of likely total costs during the construction period within, say, 95 per cent confidence limits.

Once mastered, the management of sites becomes a systematic application of the computer program which keeps the manager informed of progress against the budget and enables different contractors to see exactly where they fit into the scheme of things. Nevertheless, the manager always needs to be in a position to judge the fallible quality of the computer-generated information and then use it to make judgements. The almost infallible computer remains a mere tool of the designer or manager, dependent on the accuracy of the data fed into the machine in the first place.

Communication between different firms involved in the construction of a project can be facilitated using the Internet and E-mail. The improvement in the data-carrying capacity of cables is now common knowledge, and this innovation in telecommunications has meant that information, drawings and correspondence can be sent instantly, facilitating communications between contractors on site, head offices and even clients.

Using computers for administration

A third use of computers in the office is to aid management and improve cash flow by invoicing clients, settling bills and keeping management itself informed about the financial state of the enterprise. Programs have also been devised to help management allocate specific jobs to staff, providing a spread of the workload as well as continuity of work.

One of the main advantages of a computerised management system is that it avoids having to duplicate the same figures in different ledgers and invoices. One input will often be sufficient to place the entry in all relevant documents, thus saving clerical time and the possibility of error.

The same figures can be used to create a variety of management reports, including timesheet analyses, cost of jobs, and data on staff – to produce information on, for example, overhead costs, hourly costs, work to be invoiced and manpower requirements. The ease, speed and availability of the computer data enables management to anticipate problems and improve staff utilisation. It is also possible to integrate this information with the accounting system, to reduce the clerical workload still further.

Using computers for networking communications

One new development, reported in *Computing* (1995), has been Construction Industry Trading Electronically (CITE) which permits the companies to pass information from one firm to another. Bills of quantities and invoices can be processed between contractors and their suppliers, reducing the need for much of the paperwork. The system involves the electronic interchange of data using the rapidly expanding telephone networks in conjunction with computer developments. Eight major construction companies including Trafalgar House Construction, Balfour Beatty, Costain, Tarmac and AMEC Civil Engineering have co-operated in creating the system. The system could be used to speed up payments, reduce costs and improve communications between firms working on projects. In line with the Latham Report, it has also been claimed that the introduction of CITE will reduce construction costs significantly, but this remains to be seen.

Although management techniques exist to cope with most managerial problems, their application will always require judgement to deal with specific matters. This is especially the case where personnel problems arise, the solution of which will depend as much on the personality of the manager as the nature of the problem itself. Nevertheless, having access to a series of techniques enables a systematic start to be made to the solution, provided solutions are arrived at tactfully, by taking into account the factors that motivate the various members of the team.

Part 3
Project Appraisal

11 Introduction to Feasibility Studies

Introduction

The purpose of this chapter is to introduce the reader to feasibility studies, project appraisal, and investment analysis. Feasibility studies are an example of systems analysis. A system is a description of the relationships between the inputs of labour, machinery, materials and management procedures, both within an organisation and between an organisation and the outside world. Similarly, buildings may be seen as combinations of spaces and inputs interacting with each other and the physical, social and economic environment. Systems analysis is a set of techniques used to understand and improve methods of production. While systems analysis is usually applied to existing organisations and methods of production, feasibility studies are used to examine the practicality of proposals. In economic terms, feasibility studies are simplifying models of the real world, focusing attention on certain aspects of alternative schemes.

Feasibility studies have several uses. The most obvious purpose is to understand the cost implications of a project. The client-developer needs this information, as does the design team, in order to plan and design a building or structure. Feasibility studies set budget limits and can be used to monitor costs during construction. On completion, building users or facilities managers can use feasibility studies to monitor running costs and maintenance by comparing actual with expected figures. In theory, feasibility studies can also be used to help designers, by providing feedback on the costs and benefits of designs, components and materials. Unfortunately, in practice, it is unusual for facilities managers or designers to make full use of early project documentation, even though the figures and calculations in the documents may have been critical in coming to a decision to build.

A secondary application of feasibility studies vital to architects and developers is the use of the report to promote or defend suggested designs and proposals. In one sense, it is a marketing device for designers and developers. Apart from tender documents, the feasibility study can form the basis of a bid to carry out a project. Groups who wish to oppose schemes can also make use of the techniques of feasibility study to produce alternative solutions.

Financial versus economic viability

Feasibility studies may be used to deal with several different questions concerning the financial and economic viability of projects. Financial appraisals are concerned with cash flows, though reference may be made to non-monetary issues, but these are usually quite separate from the financial calculations. Financial viability in the private sector is concerned with the requirement that a building must generate sufficient funds to enable a developer to undertake a proposed development, repay any loans and make a profit. Banks or other financial institutions only fund projects if the proposals are expected to generate a rate of return greater than the cost of borrowing. Internal sources of finance come from within the client's organisation, while external sources of finance range from private-sector banks and investors to public-sector sources such as government-sponsored bodies and local authorities. In the public sector, profit is not the motive, and financial viability involves ensuring that sufficient funds will be available to meet the cost of providing a given public service.

Economic appraisals recognise that the total cost is not always reflected in the market prices paid for resources. The economic aspects of a construction project are concerned with the built structure's ability to meet its aims and objectives. Both financial and non-financial considerations are included in economic appraisals, whereas only financial considerations are taken into account in financial assessments. Economic viability is therefore broader than financial viability.

Judgement is always required when carrying out financial or economic appraisals, especially when deciding what to include or exclude from an economic model. The costs and benefits included depend on the questions raised and the use to which answers may be put. For instance, in financial appraisals, where the purpose of the exercise is to establish the total cost of projects, the costs relating to the site must, of course, be included. Nevertheless, a financial analysis of the cash flow would only include those items for which cash flowed into or out of an account. If the site was owned and paid for by the developer before the decision to build, the value of the land would be excluded from a cash flow.

However, if the purpose of an appraisal is to compare alternatives, only those cost elements which differ from option to option need be included. The cost of the site, for example, is clearly a major item of expenditure. However, if a project is to be built on a given site, the site costs themselves would not normally enter into a cost comparison of the designs for a building, because the identical site applies equally to all options, and the site costs of each option would cancel out.

Payments for funding are included in financial assessments of projects, but not necessarily in studies of the economics of schemes. Moreover, economic viability may deal with the problem of whether or not a project meets the requirements imposed on it by the client, the planning authorities, the local (and occasionally national and even international) community, and the final building users. Economic analyses often consider the impact of proposals on both the natural and built environment as well as the local population. A proportion of the benefits as well as some of the costs will be passed on to those living in the vicinity of any project. These gains and losses are known as *spillover effects.*

The construction of a building consists of several inputs, such as land, materials and labour. The economic value of these inputs depends on their opportunity cost. If resources are employed to create one building rather than another, then the value of these resources is the value of the next best alternative use to which they might have been put. Unemployed labour taken on to do a job on site has no opportunity cost, whereas labour taken off one site and moved to another may cause delays to the first site and as a result costing lost rental income to the building owner.

Financial feasibility studies only consider financial costs and benefits, while economic feasibility studies also take non-financial considerations into account. Economic viability therefore involves both financial aspects and broader issues, such as spillover effects and opportunity costs.

The significance of the difference between financial viability and economic viability is demonstrated in the following set of decision rules.

1. If a project is economically and financially viable, then it can proceed.
2. If it is economically but not financially viable, then the project will not go ahead, because it is anticipated that there will be insufficient inflows of money from all sources to pay for the construction, maintenance and running of the proposal, although the benefits will be greater than the costs. Unfortunately, the benefits distributed to people who are either unwilling or unable to pay. However, a project can be made financially viable through subsidies, especially if it has already been shown to be worthwhile economically.
3. If it is financially but not economically viable, then the project should not go ahead, though it may nevertheless proceed depending on the interests and responsibilities of the decision-maker. In this case, a firm may develop a site if the market is expected to generate sufficient revenues to repay loans and leave profits for the developer. Some developments go ahead because the developer gains while the costs are borne by the local population.

Legal and planning constraints

Building proposals must comply with planning constraints, and feasibility must be seen within a planning framework. An overview of all planning applications serves a useful purpose in that, although individual proposals may appear feasible, taken in conjunction with other proposals under consideration, it may be that the total mix of developments could lead to undesirable consequences.

Economists refer to the *paradox of thrift*. The paradox is that although it may be desirable for individuals to save, if everyone in the economy increases his or her savings simultaneously, demand goes down. With lower demand, unemployment increases and income declines, and lower incomes mean that people save less.

In other words, what may be of benefit to one person alone may be counter-productive if everyone were to adopt the same behaviour. Similarly, while getting from A to B in a city may be achieved most speedily by car, if everyone in a city decided to go by car at the same time, then traffic jams would result. Nevertheless, although slowed down by the congestion, it is still often quicker for an individual to travel by car than by public transport. Thus, an individual's best interest is often a compromise with others in society. The sum of the best decisions for individuals is not necessarily the most desirable outcome for the community as a whole. For this reason, the political system is drawn into decision making as an arbiter in the planning process.

Legislation may be mandatory, discretionary or enabling, depending on the importance legislators have given to the circumstances surrounding a specific situation. Since the Second World War, legislation has been concerned with town and country planning, transport, housing, the environment, and employment. Over eighty acts of Parliament have had some bearing on the production of the built environment, ranging from the New Towns Acts of 1946 and 1965, the Trees Act of 1970, the Housing Acts of 1980 and 1988, and the Town and Country Planning Acts of 1990 and 1991, to the Environmental Protection Act of 1990 and the Environment Act of 1995.

Planning affects and is affected by many aspects of government policy. Government ministries involved in the planning process include the Department of the Environment, Department of Transport, the Scottish and Welsh Offices, the Ministry of Agriculture, Fisheries and Food, and the Ministry of Defence. Certain quasi-autonomous national governmental organisations (QUANGOs) such as the Countryside Commission, the Historic Buildings and Monuments Commission and English Heritage are also involved in planning issues.

Local authorities are obliged to develop Structure Plans, concerned with employment, housing, transport, tourism, recreation and urban and rural environments. Structure Plans form a framework within which the district councils develop their own local plans in an attempt to adopt a consistent planning strategy for an area.

Before a scheme can proceed, it requires planning permission. Obtaining planning permission usually increases the value of sites and this increase is known as *planning gain*. However, before giving permission, local authorities may negotiate with developers over planning gain. In return for planning permission, developers may be required to provide a public facility such as a play area, park or library, financed out of their planning gain. This may be seen as compensation for the local community or a way of sharing the benefits of property development between the developer and the local authority.

Value for money and cost effectiveness

The purpose of studying feasibility is to obtain value for money by comparing relevant costs and benefits of different possible building options. Value for money suggests the highest return possible from a given amount of money. The concept of value for money relates objectives to costs.

Cost-effectiveness implies providing a given standard of accommodation at a minimum cost. However, it is not necessarily a matter of minimising costs. After all, the cost of construction can be reduced to zero by abandoning a proposal altogether, but this would hardly be helpful to the building users. A building must function efficiently, both in terms of the number of people using it and their activities within it. Moreover, aesthetic, environmental and social considerations must also be taken into account. Cost-effectiveness is a useful criterion to apply to the means of achieving a stated set of objectives.

At every stage of design, and even during the construction stages, designers need to consider alternative schemes, methods of construction and initial and future costs, in order to make cost comparisons and produce cost-effective designs. When variations are required by a client, comparisons can be made with the original or existing figures for the scheme, to judge whether or not the variation is an improvement or a costly change with few additional benefits.

The expected life of components also affects cash flow estimates. Moreover, during the design process, every element should be examined to see if it could be omitted without impairing the quality of the building. Otherwise, other parts of the building, in a sense, subsidise the uneconomic element. Of course, it is impossible to isolate and calculate the effect of a part of a building when the benefits are derived from the building as a whole. Nevertheless, in attempting to maximise value for money, it is necessary to identify the difference each element makes to the total building.

Cost-effectiveness is essentially qualitative and often subjective, depending on the point of view of particular individuals. In most cases, building developments are likely to benefit some people while causing others to lose. Those who lose the most may not see a profitable venture as being cost-effective from their point of view. Value for money implies that money is the major constraint, but value in terms of people or land is also important.

Life-cycle costing

Decisions to go ahead with a building project cannot be based on the initial costs of construction alone. Annual running costs including services, maintenance and repairs are also significant categories of expenditure that have to be taken into account. It is also necessary to look at how a building will function. Future revenues generated during a building's useful life, such as rental income and sales, have to be taken into account when making predictions about the future performance of buildings. External factors also impinge on the viability of building projects, including interest rates, the performance of the economy as a whole, and future prices. Government policies concerning taxation, planning and building regulations may also have a predictable impact on property valuation in the long run.

Building obsolescence describes the condition which brings about the physical neglect of structures. Obsolescence may be caused by several factors. For example, it may be more profitable to use the site for another purpose. Alternatively, a gradual decline in the fabric of a building over a number of years may make it more expensive to repair than to demolish and replace.

Economic considerations

If utility is the satisfaction or usefulness the individual derives from the consumption of goods and services, the equivalent term for a community as a whole is *welfare*. Welfare is the level of total utility of all individuals, after taking the distribution of goods and services into account. According to neoclassical economic theory, the ultimate objective in any economic decision-making process is to maximise utility for the individual, or welfare in society.

Since major building projects have consequences for society as a whole, a useful theoretical framework for considering these implications is provided by the concept of Pareto optimality, named after the Italian economist Vilfredo Pareto. Stated simply, Pareto optimality is reached when no one can be made better off without someone else being made worse off. When buildings create gains for some, they often cause losses for others. As a result, it is not possible to say that an improvement has taken place for society as a whole, since it is impossible to measure the value of one person's gain against another's loss. Only if it is possible to give to some without taking anything from others can we be certain that there must be an increase in total welfare in society, and the Pareto optimal position has still to be attained.

This principle lies behind welfare economics and provides a method of comparing the distribution of costs and benefits of proposals for change in society. If it can be demonstrated that the benefits outweigh the costs then, according to this principle, the project should go ahead. By increasing the benefits in society more than costs, there will be an increase in welfare. However, this avoids the issue of the distribution of the value of benefits and costs: for example, if the benefits are given to one person but the costs are shared among many.

Nevertheless, a project may be acceptable if at least one person gains by it without anybody else suffering a loss. This can be achieved, in theory, by compensating the losers adequately. If the construction of a building is likely to lead to a loss of amenity for some, it is possible to estimate a value of the compensation needed. The principle of compensation is that it is possible to find a sum of money equivalent to the value of a loss, on the assumption that it is possible to replace losses with other goods.

One method used to find the amount of compensation needed is the Kaldor criterion. The Kaldor criterion, named after the economist Nicholas Kaldor, may be used to compare the gains to some individuals and the losses incurred by others. To find the value of the benefits, those in favour of a particular change would be asked what the maximum would be that they would be willing to pay rather than forgo the change? The answer to this question provides the *compensation value*. Those against a proposed change would be asked what the maximum would be that they would be prepared to pay to prevent the change taking place. The answer produces the *equivalent value*. The decision would be made in favour of the higher figure. The difference between them could in theory be paid to the losers by the winners, who would still retain some of their gain.

There are, however, major practical and theoretical difficulties when comparing the welfare of one individual with that of another. For example, money is used to measure the value placed on welfare. However, money itself is subject to the law of diminishing marginal utility. In other words, the more money an individual possesses, the less a further £1 will add to his or her total satisfaction.

For this reason, £100 will not represent the same real sacrifice to a millionaire as it will to a poor person. Moreover, when placing a value on a gain, a poor individual may tend to give a smaller figure than a wealthy person, because of the difference in their attitudes towards money. Using figures given by members of the public to assess welfare can therefore lead to spurious results. Nevertheless, the interests of those affected by a project either have to be taken into account or ignored. If they are to be taken into account, the economic approach of Pareto optimality, and compensation and equivalent values provides a useful framework.

Social costs and benefits

Social costs include inconvenience to neighbours, loss of trade to competing local business, noise, extra pollution, or a drop in the value of property. The social benefits of a new conference centre in a town may include increased publicity for the town, increased trade for local shops and hotels and greater use of public transport facilities, making these less likely to be cut. From these social gains must be deducted the lost revenue from conferences which might otherwise have been held in local hotels. It must be said, however, that the creation of new facilities may also simulate interest in existing venues to the benefit of all in the area.

It is important to bear in mind that the social benefits may go to different individuals from those who bear the social costs. A factory, for example, may create employment and profits but problems of pollution, traffic congestion and inconvenience may be imposed on others. Such costs are referred to as externalities, spillover effects or intangibles. One method for weighing up losses of amenity caused by pollution, suggested by Ronald Coase, is the extension of legal rights, to enable losers to sue for compensation. The amounts settled in court would establish a value of the social costs of pollution.

Not all externalities are costs; some may be gains. When motorways are proposed, house prices in the immediate vicinity of the new road may fall, while the prices of other homes, in surrounding areas with improved access, rise. One reason property is seen as an appreciating asset is because of the spillover effects of improvements in localities with improved street lighting, roads, transport facilities, shops, schools and hospitals. As these public services develop in an area, the values of existing properties increase.

Public goods are those buildings, services or products which can be used by any individual without any further cost being incurred. Public amenities, such as parks, museums or roads are public goods. So too are television and radio broadcasts. The actual economic benefits of a public good may well extend beyond those who have paid for it, to the aptly named free-riders, who should, of course, also be included in an economic evaluation of a project.

Tangible and intangible costs and benefits

Costs and benefits should in the first instance be seen from the developer's point of view, whose main aim is usually to generate profits or obtain value for money. It is helpful to distinguish between tangible and intangible costs and benefits. In most developments, primary consideration must be given to cash flow. Thus tangible costs and benefits relate to transactions involving transfers of money.

Tangible costs include the cost of a site, construction, maintenance and service charges. Any compensation actually paid to those adversely affected is also a tangible cost, as are the costs of measures taken, for example, to counter or deter protests to new road building schemes paid for by the developer. The cost of policing may be an intangible cost to the Department of Transport, as policing is not paid for by the ministry itself. Tangible benefits include revenues from sales, rents, certain cost reductions, and savings to the developer arising from greater efficiency and improved technology.

The main difference between tangible and intangible costs and benefits is that there is a legal obligation to pay tangible costs, whereas intangible costs are not legally enforceable. It follows, therefore, that a cost or a benefit that may be classed as intangible could become tangible if action for compensation is successful. These intangible costs are then internalised: that is, converted into tangible costs when payments are made. It is therefore important, even from a cash flow point of view, to take intangibles into account. An intangible cost is a cost or part of a cost not covered by a monetary payment.

On the other hand, projects may bring benefits to those living or working near them. Nevertheless, developers rarely receive direct payment in return for improvements they make to the environment of an area. Such gains are known as *intangible benefits*. The intangible benefits of a development may be the removal of an eyesore, improved safety for pedestrians, easing of traffic congestion elsewhere, and an improvement in the status of a district.

In spite of the difficulty in assessing intangible costs and benefits accurately, some valuation of these gains and losses is useful in providing an overall picture of the impact of a development on its local environment and community. Intangible costs and benefits cannot be ignored, because they are implied when decisions to invest are taken. For example, the average value per employee placed by employers on their office environment is implied in the budget used to maintain it. This can be calculated by dividing expenditure by the number of individuals involved. Thus, if the aim of a particular expenditure of £10 000 per annum is to enhance the environment of 1000 employees, then the implicit assumption is that the benefit per individual is worth £10 per year. Of course, such an analysis is an over-simplification, since any improvement to an office will have several beneficial consequences, including, perhaps, raising staff morale, reducing staff turnover and therefore lowering training costs and increasing productivity. Other intangible benefits of a suitable working environment may be measured by considering the value of reductions in errors by staff and the reduction in the amount of sick leave taken. Factors such as these may be included in an assessment of the costs and benefits of a proposal, although they lie outside the scope of pure financial analysis.

However, some method has to be found to evaluate these factors, in order to weight the arguments in favour of or against a project. Interested parties who wish to prevent the construction of a given project may delay or force changes on proposals. In order to minimise delays and the risk to the developer of wasting expensive preparatory work, it is important to anticipate the strength of feeling of objectors and attach a sufficient value weighting to their intangible costs. Objectors are often members of the local community most affected. Issues which may appear to be of minor importance to the developer can turn out to be insuperable stumbling blocks at a public enquiry. Of course, it is true that on occasion apparently minor issues given political support can be exaggerated to frustrate projects. However, the planning system is designed to preserve the rights of individuals in a democratic process and to intervene on behalf of society at large.

Note on double counting

One of the pitfalls to avoid in any accounting exercise, especially one involving intangible costs and benefits, is the problem of double counting, which is the inclusion of the same element twice. On the second occasion it may appear in a different form in the accounts. For example, a new motorway link may save travelling time for commuters. The time saved can be assessed and given a value weighting in order to measure this particular social benefit.

However, double counting would occur if an increase in property values were then also included in the list of benefits. If property prices rise as a consequence of the new road, it is precisely because travelling time has been reduced. The increase in the price of property includes a valuation of the travel time saved. Only one or the other of these factors may be included in a appraisal.

Weighting intangibles using money

Shadow pricing is one method of weighting intangible costs and benefits using money as the weight. The more important the intangible cost, the greater its shadow price and the more significance it has on the costing of the project. By using shadow prices, intangible costs and benefits can easily be combined with the more conventional financial costs and revenues in order to calculate the overall economic viability of a scheme.

Shadow pricing involves finding and attributing values to costs and benefits even when no money changes hands. There are a variety of approaches to evaluating intangible costs and benefits. For example, devising contour lines around a proposed site in order to map the intensity of impact of a scheme can help to distinguish between those most and least affected. Figure 11.1 illustrates three zones, A, B and C, in order of the diminishing impact of a proposed scheme in zone A. Presumably, those living or working beyond zone C, in D, would not be affected by the proposal. Once the contour lines have been drawn on a map of an area, like the slope of a hill, the level of potential inconvenience or disadvantage can be seen to rise in bands according to the location of the lines as one approached the site of the scheme in zone A. For example, as one moves nearer to the flight path of an airport, the greater is the disturbance caused by the noise of aircraft landing and taking off. But this

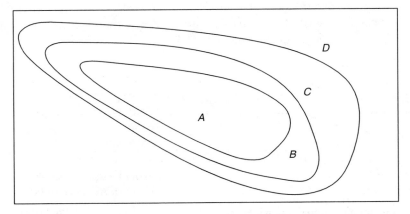

Figure 11.1   Contour lines showing diminishing geographical impact of proposal

still leaves the problem of attributing money values as compensation for the intangible cost of noise pollution.

It may be necessary to evaluate intangible costs, especially those likely to remain during the life of the project and beyond. For example, a proposal to build an industrial estate implies a number of intangible costs, such as the effect of increased noise levels, the presence of heavy goods vehicles, and the possible deterioration of the environment. Some households will be more concerned than others, usually the nearer they are to a proposed site. The number of people adversely affected can thus be estimated by using the contour lines on a map.

Table 11.1 illustrates this procedure showing the percentiles of the population in descending order of intensity, according to money weighting. Assuming a stable total population of 20 000 people in a given location, the percentiles, or proportions, of the population affected by a proposed scheme might be estimated as follows: the percentile most affected in zone A might be 10 per cent of the population; the next percentile, in zone B, may be 20 per cent of the population; while the remaining 70 per cent of the population in zone C might be only affected minimally. These percentages can then be used to weight the valuation of the intangible cost.

At this point, a survey might be used to find values to attach to social costs. Unfortunately, very often it is not possible to conduct a survey in a sensitive area because of the nature of the development process. Arousing suspicion and hostility to possible developments would be counter-productive from the developer's point of view. Nevertheless, a monetary weighting may be arrived at by using techniques similar to the Kaldor criterion or, more likely, by negotiation with those making the investment decision. If the values given to intangible costs are to be useful, it is vitally important that the economic adviser remains independent of those directly involved, such as the developer or the local population. It may be necessary simply to impute costs; that is, to allow for a cost *as if* money had been paid.

Table 11.1   The computation of an intangible cost

Total population	Percentile	Group size	Money weighting	£
20 000	10%	2 000	£50 per head	100 000
20 000	20%	4 000	£20 per head	80 000
20 000	70%	14 000	£10 per head	140 000
Total valuation of annual intangible cost				**320 000**

Assume the result of the survey is as shown in Table 11.1. The estimate of the intangible cost to the first percentile is valued at £50 per head per annum, the next group at £20 per head per annum, and the last group at £10 per head per annum. Although the method may appear to be arbitrary, this exercise demonstrates a set of assumptions and valuations needed for a project to be adopted or rejected. If it can be argued and agreed that the intangible costs exceed £320 000 per annum, then the project should be rejected.

Similar procedures to those used to establish intangible costs can be applied to evaluate intangible benefits. The value of a benefit is frequently higher than the price charged, because the price only reflects the interaction of the market forces of supply and demand. Consumer surplus is the difference between the amount an individual is willing to pay and the amount actually paid. An allowance for consumer surplus gives a more accurate value of a benefit than does the price. A survey of potential users would reveal the amount they would be willing to pay for a given benefit. One method of estimating the consumer surplus is to find the compensation and equivalent values of the Kaldor criterion. It is clear that the answers to the two questions would not produce the same results, but a combination of the two may provide an estimate of consumer surplus. An individual's valuation of a benefit will depend on his or her circumstances, income and tastes. Survey results could be used to represent a sample of the population to establish average values. Consumer surplus can be found by subtracting the actual price from the result of the survey.

If one were attempting to value the benefits of a museum which did not charge an entrance fee, a value for each visit by members of the general public could be estimated. The estimate could be based partly on an equivalent entrance charge, its shadow cost, and partly on an estimation of the consumer surplus. The consumer surplus applies to the amount over and above the shadow price. Therefore, there is no question of double counting.

An intangible benefit of a museum or other tourist attraction may be the extra visitors drawn into an area, who then go on to spend money in local shops. An assessment of the value of this social gain may be based on a survey of the local retailing facilities and their potential appeal to visitors, together with an estimate of the amount of extra spending that might be expected.

There will also be specific social costs as a result of building an amenity such as a museum. For example, a number of cars coming into an area would require extra parking facilities. The cost of providing off-street parking spaces may be included as one of the tangible costs of the museum project, if, in fact, they are to be constructed. However, if on-street parking or present parking facilities are to be used, the extra vehicles would deprive current users of space. In those circumstances, an intangible social cost would be passed on to others.

To estimate the cost of parking the extra cars, the extra time spent looking for a parking space, the number of vehicles and the cost per vehicle would have to be assessed. A survey could be used to approximate the shadow costs of parking, by estimating the proportion of visitors who come by car and establishing the value of a parking space both from the point of view of visitors as well as the local residents, who would be displaced.

Valuation of human life in economic terms

Various approaches to evaluating intangible costs and benefits attempt to place a monetary figure on factors which influence decisions to build. One of the most difficult questions raised by attempts to cost intangibles is the valuation of human life. Investment decisions often contain implicit assumptions about the value of life. Placing a monetary value on life is emphatically not to reduce actual human worth to figures but to state in simple terms the assumptions implicit in a decision.

There are several possible methods for calculating the value of human life. One method is to assume that a human life is equivalent to 6000 working days. In the case of professionals, one court ruling was the earning power over the remainder of the person's working life.

Alternatively, valuation may be based on the amount which might be spent to prevent accidental death. The implied value of life is the probability of a death multiplied by the amount spent to prevent fatalities occurring. A safety measure budget of, say, £55 000 implies a value of human life, if the budget allocated for safety measures is carried out solely in order to prevent a notional death caused by a failure of part of the building, such as collapsing masonry. If the likelihood of a fatal accident occurring during the life of the building is assumed to be 1 in 1000, the implied value of a human life in this case would be £55 000 × 1000, namely £55 000 000. Another method of evaluating human life uses statistics from various sources, such as the average compensation currently awarded by courts, or the average value of recent independent analyses.

The implicit value of human life can also be calculated by estimating the amount spent on various recent projects in order to reduce the loss of life. However, this method poses difficulties in the choice of projects used to base an average upon. The amount spent varies, depending on the nature of the project. In one situation, such as in a factory, life may be saved by erecting a warning sign, while in another situation at the other extreme, many millions of dollars are spent by the US Space Agency at NASA in attempting to make travel safer for the few astronauts brave enough to venture out of the earth's atmosphere.

An implied valuation of human life is assumed in the appraisal of hospital building projects, especially those hospitals in the public sector, which are unlikely to be viable in financial terms. Each hospital will be designed to treat a given number of patients, some of whom will be saved because of the new facilities offered by the new building. To simplify the method for the purpose of illustration, assume:

1. There are no intangibles other than the number of expected human lives saved.
2. The expected life of the proposed hospital is 10 years.

The minimum implicit value placed on saving a patient's life is the average value necessary to make the total of financial and intangible benefits equivalent to total financial costs, at the required discount rate. Thus, if the financial costs exceed financial benefits by £5m, and assuming 1000 additional patients are likely to be saved during the life of the building, then the present value of each patient's life is, in effect, worth £5000. This is the required figure per patient needed to make the proposal economically viable. The implicit valuation of a life is the excess of discounted costs over discounted benefits divided by the expected number of extra lives saved. Thus:

$$V = \frac{c - b}{n} \qquad\qquad (11.1)$$

where

$V =$ implied value of a human life
$c =$ discounted tangible costs
$b =$ discounted tangible benefits
$n =$ expected number of extra lives saved

and

$c > b$

It can then be argued that unless 1000 extra patients can be saved, at a value of £5000 per head, then the project should be rejected. That is, if it is not deemed to be worth £5000 per patient, the hospital should not be built, as the cost of the hospital is £5m more than its revenues. This is not to say that a human life is only worth £5000. This value per patient is simply implied by the assumptions in the economic model of the costs and benefits if the hospital is built.

Because of the apparent arbitrariness and variety of approaches to evaluating human life, it is not surprising that the methods and results have been rejected by many on moral as well as economic grounds. An alternative is to make intuitive decisions by avoiding the issue. The problem may be dealt with by simply stating the number of people likely to be saved or killed by a given project. One method used frequently which avoids placing a monetary value on a variety of intangible costs and benefits is importance-weighted effectiveness rating.

Importance ratings

Many economists reject the techniques of shadow pricing and the idea that one can attach meaningful money values to intangibles. They argue in favour of placing intangible elements in order of priority, using importance rating, which is a points system. A simple method for rating the relative effectiveness of each alternative option is illustrated in Table 11.2. There are five steps, at the end of which each proposal is given a rated value, which can be compared with others.

Table 11.2   *Five steps to points system of evaluation*

Step no.	Activity
1	State the objectives of the project in order of importance.
2	Give the first priority a value of 100 and the other priorities a relative rating.
3	Give a percentage mark to show how effective the proposal is in achieving a stated objective.
4	Multiply the importance rating percent by the effectiveness mark, to give a weighted effectiveness rating.
5	Sum the weighted effectiveness ratings. This is the rated value of the proposal and may be compared to the alternatives by using the same table and method.

The first step in carrying out a feasibility study should be to generate the aims of the project (Gruneberg and Weight, 1990). The aims and objectives of a construction project can be defined in terms of three elements. These are:

1.   Function – this can be defined by the activities to be undertaken within the space of a building.

2. Quantity – this can be defined by the number of users in each space, for example, pupils per room, patients per ward, people per flat, guests per hotel, cars per hour, passengers per day, volume of water per hour, and so on.
3. Quality – this can be communicated in lay terms using, for example, a similar building, to illustrate the quality of the finishes to be expected: 'a four-star hotel like. . .', or 'a theatre like . . .'.

Each objective should begin with the word, to . Thus a set of objectives for a hotel might be:

(i)   To hold conferences for up to 250 delegates with facilities equivalent to those in the ABC Conference Centre.
(ii) To provide quality goods for sale in six small specialist retail outlets in the concourse.
(iii) To accommodate 600 guests per night in four-star quality.
(iv) To provide 200 covers per evening in a restaurant, similar in quality to the restaurant in Hotel XYZ.
(v)   To provide sports facilities for up to 20 guests at any one time, similar to the 321 Sports Centre.
(vi) To have a bistro-style restaurant providing up to 500 meals per day, similar in quality to Café 123.

It is worth noting the difference between constraints and objectives. Whatever is chosen must comply with a constraint, while an objective simply defines the purpose. A constraint can be seen as a negative criterion, because a project cannot go ahead if it does not lie within the planning, financial, technical or environmental limits set. Objectives, on the other hand, can be viewed as positive criteria, since a scheme may go beyond its original objectives provided these are satisfied. Indeed, if a project surpasses expectations set by the objectives, it would generally be viewed as a success and a bonus for the developer.

Having established the various objectives, the first step is completed by placing the objectives in order of importance. Only the four most important objectives have been included in order of importance in Table 11.3, and steps 2 to 5 in Table 11.2 are then followed. Thus the accommodation of 600 guests is given a weighting of 100, and Option A is deemed to have scored 70 per cent in terms of the quality of the accommodation proposed. Although Option A provides a higher perceived quality for its conference centre, the clients have not deemed this to be as important as either hotel accommodation and quality restaurant facilities.

The same importance-weighting is used against each option and is applied to each option's effectiveness mark. The importance-weighted effectiveness ratings of four options are shown in Table 11.4.

The table shows that although Option A has the highest score in aggregate, Option D is seen as having the best score for the quality of its accommodation. Option C has the best restaurant and conference facilities. Even Option B, which has the second lowest score, is given the same score for its sports centre as Option A.

Table 11.3   Evaluation worksheet

Objectives	Option A Importance rating	Effectiveness mark (%)	Weighted effectiveness
To accommodate 600 guests	100	70	7 000
To provide 300 restaurant covers per evening	90	65	5 850
To hold conferences for up to 250 delegates	40	80	3 200
To provide sports facilities for up to 20 guests	20	60	1 200
Project rating			**17 250**

Table 11.4   Comparison of options

Objectives	Option A	Option B	Option C	Option D
To accommodate 600 guests	7 000	6 800	*5 500*	**7 300**
To provide 300 restaurant covers per evening	5 850	4 950	**6 300**	*3 600*
To hold conferences for up to 250 delegates	3 200	*1 600*	**3 400**	2 400
To provide sports facilities for up to 20 guests	**1 200**	**1 200**	860	*700*
Total	17 250	14 550	16 060	14 000

Note: Figures in bold highlight preferred options on particular criteria. Figures in italics emphasise the option with the lowest assessment on particular criteria.

A similar approach can be taken in assessing the relative merits of different proposals for a civic theatre, a hospital or a road. The objectives for a theatre project may include providing a facility to put on new plays, improving the cultural environment and appeal of a town to its citizens, promoting a city with good PR to attract new industry and jobs, and ensuring the town is given the status symbol of an attractive theatre.

Importance-weighted effectiveness rating is a systematic approach allowing all options to be examined and compared on a number of criteria. The figures are not as important as the process of arriving at them. Indeed, it may well be that Option C is selected ultimately, in spite of its disadvantages because of constraints of a planning or financial nature. In any case, each option has advantages and disadvantages compared to its alternatives. The final choice is often a matter of judgement.

This method separates the intangible costs and benefits from the financial considerations of a scheme. The main advantage of this approach is that it simplifies the many real problems of attributing prices to intangibles, such as the desirability of offering sporting facilities to only a very small proportion of visitors to the hotel. For this reason it can be argued that it is more complete than those which incorporate both financial and non-financial costing techniques within one mathematical economic model. A mathematical model including the intangibles may make assumptions about relationships between variables that are not always stated. By separating the intangible costs and benefits from the financial considerations of a proposal it allows a clear statement of the objectives of a project. These can then be achieved by using the most cost-effective method.

The disadvantage of this method is that it separates the intangibles from the tangibles. It is therefore not possible to weight all the costs and benefits of a project in a single mathematical model combining all the elements and it may lead to excess expenditure on projects for which the implicit values of the intangibles have not been assessed.

Although it can be argued that weighting objectives in this way is an oversimplification of the relationship between the different variables and is largely invalid, the values, weightings and relationships are not intended to be viewed as predictions about the future. This method, as with alternative strategies for measuring intangibles, simply presents a set of assumptions which help to throw light on decisions that have to be taken. Rating projects can be a useful tool for decision-makers. It is not a substitute for judgement.

Since feasibility studies can only be concerned with options that are possible, legal constraints must always be taken into account and complied with. Planning restrictions therefore determine the limits of a project. Other legal or formal constraints include the tax system. Taxes, such as Value Added

Tax (VAT) and Corporation Tax, may have consequences on the choice of materials and finishes. Given these constraints, methods, using economic criteria, including those described in this chapter need to be found for evaluating the gains for the gainers and the losses of the losers of any scheme, before a proposal can be judged to be economically viable or not.

12 Financial Appraisal

Introduction

The previous chapter introduced the concept of feasibility studies as an example of systems analysis and economic modelling. This chapter discusses feasibility studies as an example of investment appraisal applied to construction projects. Similar techniques are used in a variety of different contexts, such as investment in new plant and equipment, or in shares in the stock market. The basic questions facing every investor are whether or not the investment will be worth more than the amount invested, and if so, by how much? Alternatively, an investor may need to know the probable rate of return on an investment to compare it to the rate of interest to be paid on loans to finance the project.

In order to answer these questions there are available a number of financial techniques, which use discounting. However, because of the simplifying assumptions, uncertainty and unavoidable errors of feasibility study techniques it is often advisable to treat the mathematical calculations with a healthy degree of scepticism.

The principle of discounting

A financial investment is an expenditure in one period with a view to receiving a flow of money in the future. The problem is that money in the future is worth less than money in the present, even without inflation! Without inflation money would still purchase the same quantity of goods at some time in the future, but the delay in purchasing has to be taken into account. Is it better to have £1000 today or £1000 in ten years? Money today can be used to purchase goods or deposited in a bank to earn interest, whereas there is always a risk that one may not be alive in ten years hence. The question is, how much is one willing to pay now in order to receive £1000 after waiting ten years? The answer is usually less than £1000. It is therefore not meaningful to add the costs of construction to the annual costs of repair and maintenance over the life of a building because the figures arise in different years and are not of equivalent value.

The cost of capital to a borrower is the interest rate, which is the rate at which an amount of money grows over time and needs to be repaid. If the interest rate is 10 per cent, then £100 today will be worth £110 one year from now and £121 in two years' time. Compound interest converts a present day amount of money into a future value. A discount rate is the same as a compound interest rate except that, whereas compounding moves from the present to the future, discounting moves from the future back to the present. Discounting calculates the present value of a future amount. For example, £110 in one year's time is worth £100 today, assuming a discount rate of 10 per cent. As future costs and revenues are not equivalent in value with present prices, expected future figures must first be discounted.

The time value of money may be seen as the value of the delay before receiving payment. The greater the value of the delay, the higher the interest rate needed to compensate the lender or attract the investor. The time value of money varies from individual to individual, depending on age, personal requirements and personality.

The future value of a sum of money using compound interest is given by Equation (12.1):

$$PV(1 + r)^I = FV \qquad\qquad (12.1)$$

where

PV = the present value
FV = some future value
r = the interest rate
I = number of periods hence.

The formula used for discounting is derived by transposing terms:

$$PV = \frac{FV}{(1 + r)^I} \qquad\qquad (12.2)$$

where

r = the discount rate.

While it is always useful to understand the principles that lie behind techniques, discounting is invariably included in spreadsheet and accounting programs. Financial modelling involves using computers to calculate present values at different discount rates. Normally, cash flows should extend to the expected life of a building (say fifty to sixty years), but the present values of future costs and benefits in forty or even thirty years' time tend to make little difference to investment decisions, since their discounted present values are usually relatively small, depending on the discount rate used.

Net discounted present value

The net discounted present value (NDPV) is used to answer the basic investment question of how much more or less an investment is worth compared to the amount invested. Many firms see investment as a way of increasing the value of their assets. The NDPV shows the present value of an asset. The value of an asset (or a proposal) is derived from the stream of revenues, less costs, generated in the future. Annual net revenues before tax are revenues less costs in any given year. Each year's net revenues must first be discounted to the present in order to make all values equivalent. The initial costs of construction are then subtracted from the discounted net revenues to produce the NDPV. The higher the NDPV, the greater the increase in the value of assets. Thus, if the NDPV in the feasibility study of a building project is £500 000 (at a given discount rate), this implies that the building will in the long run generate net benefits which, when discounted back, will be equal to £500 000 more than the cost of the investment.

It is necessary to make the following assumptions;

1. All future figures are known with certainty.
2. A given building costs £900 000 to complete.
3. On disposal after five years, the building will have appreciated in value by 10 per cent.
4. The annual running costs will be £50 000.
5. The annual rental income will be £300 000.
6. All payments are made at the end of each year.
7. There is no inflation.

Table 12.1 illustrates the method. The number of years is given in column 1. Net benefits are costs less benefits and three discount rates have been used, 15 per cent, 20 per cent, and 25 per cent.

In table 12.1, the NDPV can be seen to vary depending on the rate of discount. Using a discount rate of, say 20 per cent, produces an NDPV of £66 615. The higher the rate, the lower the NDPV. This occurs because, at high rates of discount, future gains are less significant. The present value of a future sum of money depends on the discount rate chosen, which in turn depends on the cost of capital, the anticipated life of a building itself, its future use and the level of risk associated with a given project.

Different rates of discount are used for different strategic purposes and different types of decision. For example, some feasibility studies are carried out to demonstrate the extent of financial exposure to risk by showing relatively high rates of return. Low rates of discount of between 5 and 10 per cent

Table 12. 1   Discounted cash flow table, calculating NDPV with payments in arrears

Year	Cost £	Revenues £	Revenues less costs £	5(£)	Discount rate (%) 10(£)	15(£)	20(£)	25(£)
0	0	0	0	0	0	0	0	0
1	900 000	0	−900 000	−857 143	−818 182	−782 609	−750 000	−720 000
2	50 000	300 000	250 000	226 757	206 612	189 036	173 611	160 000
3	50 000	300 000	250 000	215 959	187 829	164 379	144 676	128 000
4	50 000	300 000	250 000	205 676	170 753	142 938	120 563	102 400
5	50 000	990 000	940 000	736 515	583 666	467 346	377 765	308 019
			NDPV	**527 764**	**330 678**	**181 091**	**66 615**	**−21 581**

Note: Year 0 is the present year. As payments are not made until the end of the 'present' year, zero payments appear in all columns in Year 0.

are used to evaluate the life cycle costs of buildings. Low discount rates are also used to highlight the long-run benefits and considerations of public-sector schemes. Indeed there is even a strong case for using a discount rate of zero for decisions with very long-term implications, such as the problems of decommissioning nuclear power stations.

Low rates of return on property investment is not unusual. In the late 1980s yields dropped to as little as 2 per cent as buildings appreciated in value, and the yield on the capital invested was only a part of the overall return expected from an investment, the other part being the realisation of a speculative gain when the building was eventually sold. Yields are dealt with later in this chapter.

Table 12.2 illustrates a discounted cash flow with payments in advance of each period. Since all payments are made earlier than those shown in Table 12.1, the net present value using the same discount rate of 20 per cent increases to £79 938.

Table 12.2   Discounted cash flow table: calculating NDPV with payments in advance

Year	Cost £	Revenues £	Revenues less costs (£)	5(£)	Discount rate (%) 10(£)	15(£)	20(£)	25(£)
1	900 000	−	−900 000	−900 000	−900 000	−900 000	−900 000	−900 000
2	50 000	300 000	250 000	238 095	227 273	217 391	208 333	200 000
3	50 000	300 000	250 000	226 757	206 612	189 036	173 611	160 000
4	50 000	300 000	250 000	215 959	187 829	164 379	144 676	128 000
5	50 000	990 000	940 000	773 340	642 033	537 448	453 318	385 024
			NDPV	**554 151**	**363 747**	**208 254**	**79 938**	**−26 976**

The internal rate of return

The internal rate of return (IRR) is the percentage rate of return of capital invested in a project. It would not be feasible commercially to borrow at a rate of interest higher than the IRR.

Construction projects are invariably financed by a number of parties including banks, insurance companies, pension funds and the developers themselves. Even when developers partly finance their own projects, the interest payments to themselves must be imputed; that is, treated as if paid to an outside body. This ensures that all the financial costs of a development are taken into account and covered by the project. Whether a project is financed internally or by an outside institution, interest payments are included by deducting interest from each year's net revenues before discounting. However, it is possible to omit interest payments from the calculations, especially if the interest rate is not yet known or agreed with the lender, in order to find the internal rate of return before interest.

Like the NDPV, the IRR can be found on all spreadsheet programs. However, a simple method of estimating the IRR can be used by calculating the NDPVs of a project at various rates of discount. Three discount rates are usually sufficient. The net discounted present values at discount rates of, say, 15, 20 and 25 per cent are plotted on a graph and a curve is drawn as shown in Figure 12.1.

The IRR is given at the point of intersection of the curve and the horizontal axis. This represents the rate of discount which would generate no increase or decrease in the value of the assets, where total discounted costs are equal to

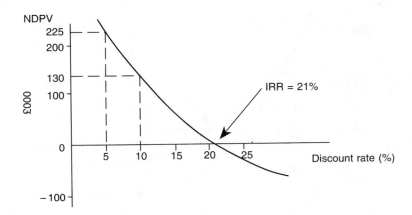

Figure 12.1   The internal rate of return

total discounted revenues. The IRR remains the same whichever rate of discount is used to establish the net present value. This can be seen in Figure 12.1, which shows that a discount rate of 5 per cent produces a net present value of £225 000, and a discount rate of 10 per cent produces a net present value of £130 000.

One weakness of the IRR criterion is that the option with the highest IRR is not necessarily the best option to choose in all circumstances. In Figure 12.2, two options are plotted. Option B has the higher IRR (21 per cent). However, at rates of discount below 16 per cent, Option A has higher net present values. For example, at a discount rate of 15 per cent, the NDPV of Option B is only £200 000, whereas Option A is approximately £230 000. If the cost of borrowing is below 16 per cent, Option A would be preferred, but Option B would be selected if the cost is above 16 per cent. At a target discount rate of, say, 18 per cent, the NDPV of Option B is £80 000, compared to the NDPV of Option A, of approximately £40 000.

The gradients of the two options reflect the fact that options with relatively high slopes and low IRRs tend to have higher capital costs with lower running costs. Projects with high capital outlays and low running costs are only financially viable when interest rates are low. When discount rates are high, the discounted present value of future benefits does not compensate for the capital outlay. The IRR, used as an investment criterion, tends to favour those projects with relatively low costs in the initial years and with relatively high running costs in the distant future, whereas it may be desirable to have a higher capital outlay and a more durable, reliable, energy efficient building.

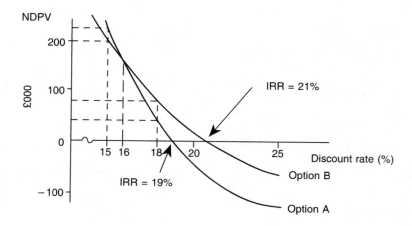

Figure 12.2   Comparing the internal rates of return of two options

However, this is not always the case. Where buildings are liable to be affected by changes in fashion or taste, such as retailing, restaurants and even office accommodation, there may be little point in having finishes which become obsolete long before they require replacement.

Annual equivalence and capital values

It is sometimes useful to compare the implied costs or revenues of proposals on an annual basis. Annual equivalence can be used for comparisons if, for example, one option involves large capital expenditure and low annual running costs, while an alternative proposes low capital expenditure but high costs of repair and maintenance.

An annuity is an arrangement entitling the investor to a regular amount of money, usually paid annually, in return for an initial payment of a capital sum. The annual equivalence is the value of annual payments out of a capital sum. Any sum of money can be seen in terms of either its capital value or the amount it would generate annually. Both figures are equivalent.

The size of the annual equivalence depends on the yield. Yield is the rate of return usually expressed as a percentage. As shown in Equation (12.3), it is the ratio of expected annual revenue (after costs have been deducted), to capital:

$$r = \frac{(R - C)}{K} \tag{12.3}$$

where

r = investment yield
R = annual revenue
C = annual costs
K = capital invested

Transposing terms, Equation (12.3) becomes:

$$R - C = K \times r \tag{12.4}$$

As we wish to know the amount of capital equivalent to an annual sum,

$$\text{let} \quad A = R - C \tag{12.5}$$

where A = the annual equivalence.

The annual equivalence of £100 at a yield of 10 per cent is £10 per annum. The relationship can be seen in the formula:

$$A = K \times r \qquad (12.6)$$

If $K = £100$
and $r = 10\%$.
Substituting in Equation (12.6),

$$A = £100 \times 10\% \qquad (12.7)$$
$$= £10.$$

Transformation of Equation (12.6) gives the equivalent capital value of an annual sum:

$$K = \frac{A}{r} \qquad (12.8)$$

or

$$K = A \times \frac{1}{r} \qquad (12.9)$$

The inverse of the yield, $1/r$, is called the *years' purchase*. The capital value of a property, K, is found by multiplying the annual rent less expenses by the years' purchase factor. For example,

if   the property yield $= 10\%$
and   annual rent less expenses $= £100\,000$

then substituting in Equation (12.9)

$$K = \frac{£100\,000}{1} \times \frac{1}{1/10} \qquad (12.10)$$
$$= £100\,000 \times 10 \qquad (12.11)$$
$$= £1\,000\,000 \qquad (12.12)$$

The benefit:cost ratio

The benefit:cost ratio enables a comparison to be made between projects of different durations and size. It relates the size of the benefits to the cost of acquiring them. The formula for the benefit:cost ratio is:

$$\text{Benefit:cost ratio} = \frac{\text{Sum of discounted benefits}}{\text{Sum of discounted costs}} \qquad (12.13)$$

The higher the ratio, the better the project. This ratio can be used to confirm decisions. However, the tendency is towards lower benefit:cost ratios, the larger the project.

Payback periods

The simple payback period is probably the most commonly used and easily understood investment criterion. It is a measure of risk, since it measures the number of years it will take to recoup investment in a project. Any capital expenditure involves the outlay of funds in the initial period followed by a flow of income or reduced costs. The payback period is the length of time required to equate the original expenditure and the resulting income or flow of cost reductions. Thus, if a building cost £10m and expected net rental income is £2m per annum, income will be equal to the capital outlay after 5 years, which is the payback period. The formula for the payback period is:

$$\text{Payback period} = \frac{\text{Capital outlay}}{\text{Annual net income}} \qquad (12.14)$$

The main advantage of the payback period calculation is its simplicity. It uses actual cash flow expenditures and receipts. It may also confirm the superiority of an option, in so far as it indicates one important aspect of the risk associated with projects. Namely, the longer the payback period, the longer the client is exposed to the danger of not recouping the original investment through unforeseen circumstances. If funding was required for a second project at some point in the future, the payback period may be used to ensure that funds would be available on time.

One of the main objections to the use of the payback period criterion is that it disregards the need to discount future costs and income. However, a discounted cash flow can also be used to adjust the payback period, using the cumulative present values of future net benefits. The cumulative discounted payback period is found by combining each year's discounted cash flow until the cumulative cash flow becomes positive. In the initial phases of the life-cycle of a project, the cash flow is negative because of the costs of site acquisition and construction, but once a project starts to generate revenues, the annual cash flow becomes positive. These positive annual discounted cash flows accumulate and eventually equal the original investment.

Although the net benefits continue to accumulate after the payback period, succeeding years are ignored, even if major expenses are incurred in the following year, or if the project then collapses. This is a major disadvantage of both the payback period investment criteria using actual cash flows or cumulative discounted net benefits. It ignores the expected useful life of a project, which is especially significant when dealing with property matters, where buildings frequently have expected useful lives of several decades. Establishing the payback period does not help in finding those options which maximise profits or benefits. It is not possible to tell from the payback period which projects have the highest NDPVs or IRRs.

Nevertheless, payback periods are often used by managers in industry, who hold their positions for relatively short periods of time before moving up their organisational ladder or gaining promotion elsewhere. Projects with short payback periods help to demonstrate managers' ability and show off their achievements before they move on.

To overcome the objection that payback does not take the time value of money into account, the discounted payback criterion is sometimes used. This takes the present value of costs and future net revenues and finds the year when the cumulative discounted net benefits are equivalent to the capital invested. As might be expected, the cumulative discounted payback period is greater than the simple payback period calculation.

Analysis of cost effectiveness

In practice, by the time a client approaches the construction industry, the objectives have already been decided. In that case, the main purpose of option appraisal is to find the most cost-effective way of achieving the aims set by the developer. To do this, only the costs of various options need to be considered. All costs need to be taken into account and discounted. Table 12.3 includes construction costs, costs in use and residual valuation after 10 years.

Table 12.4 shows the present value of each year's net cash flow derived from Table 12.3. The net discounted present value is the sum of the discounted net annual cash flows. Thus, given any of the discount rates in Table 12.4, the present value may be stated as any one of the following:

> Present value @ 5% discount rate = −£8 259 000
> Present value @ 10% discount rate = −£8 671 000
> Present value @ 15% discount rate = −£8 784 000.

The NDPVs are all negative because this project is not intended as a generator of revenue. In such schemes it is usual practice to consider cost-effectiveness in terms of cost yardsticks. A cost yardstick is a cost per unit. A unit is the most appropriate measure in any given project. In a school, the unit of measurement may be the number of pupils, in a hospital it may be the number of bed spaces or the number of patients, in a housing project it may

Table 12.3 Discounted cash flow of one option (£000s)*

Year	0	1	2	3	4	5	6	7	8	9
Expenditure										
Building works[a]	6 190									
L.A. tax[b]		92	102	112	123	135	148	164	180	198
Staff costs[c]		490	527	566	609	655	703	757	813	874
Security, other staff[c]		20	22	23	25	27	29	31	33	36
Light and heat[d]		20	22	22	24	25	27	28	30	32
Maintenance[e]		5	6	6	12	7	8	9	17	11
Finance costs[f]		290	290	290	290	290	290	290	290	290
Total	**6 190**	**917**	**969**	**1 019**	**1 083**	**1 139**	**1 205**	**1 279**	**1 363**	**1 441**
Residual value[g]										(14 565)
Less										
Nominal outstanding loan										5 250
Expenditure in Yr 10										1 441
										(7 874)

Notes:
* This example is based on the work of Ben Smith, postgraduate architecture student at the Polytechnic of North London, 1987.
(a) Prices given refer to a specific date.
(b) Annual rate of increase in local authority taxes assumed at 10 per cent per annum.
(c) Wage inflation rate 7.5 per cent per annum.
(d) Energy cost inflation rate assumed at 6 per cent per annum.
(e) Maintenance cost inflation at 10 per cent per annum, with redecoration every four years.
(f) Figure includes repayment of principal and interest. Treasury borrowing rate of 5 per cent per annum on 30-year loan.
(g) Based on current rent of £11 per ft^2, (£118 per m^2) annual growth rate at 15 per cent.
(h) Years' purchase is assumed to be 7.5.
(i) The figures in brackets refer to revenue items and are deducted from costs.

Table 12.4 Present value of each year's net cash flow

	Discount rates (%)		
Year	5	10	15
1	−6 190	−6 190	−6 190
2	−873	−834	−797
3	−879	−801	−733
4	−880	−766	−670
5	−891	−740	−619
6	−892	−707	−566
7	−899	−680	−521
8	−909	−656	−481
9	−922	−636	−446
10	5 076	3 339	2 239
NDPV	**−8 259**	**−8 671**	**−8 784**

Table 12.5   *Summary of expected costs and benefits and implied investment criteria:*
time horizon; 60 years; target discount rate; 15%; yield; 10%

Investment criteria	Option 1	Option 2	Option 3
Construction costs (initial)	**£10m**	**£15m**	**£12m**
Expected annual running costs	£2.5m	£3m	**£2m**
Expected annual benefits	**£4m**	£5m	£4.5m
Net discounted present value @ 15%	£ 0.26m	£ 2.02m	**£4.23m**
Internal rate of return	14.6%	12.9%	**20.4%**
Annual equivalence, yield = 10%	£3.408m	**£4.267m**	£1.208m
Cost–benefit ratio @ 10%	1.13	0.91	**1.39**
Discounted cumulative payback @ 10%	12 years	22 years	**6 years**
Payback	6.7 years	7.5 years	**4.8 years**

be the number of bedrooms, and in a road scheme it may be the number of vehicles per hour or per week.

A comparison of the discounted present values of the costs of different options provides the most cost-effective proposal. Although the cheapest solution may be found in this way, other considerations may be of overriding importance. Making the choice on the basis of an alternative criterion does not invalidate this method of cost comparison, since it will always be relevant to know the excess over the cheapest option that the client must bear in order to have the advantages of a preferred scheme. If alternative options are being considered for, say, the same site, then there is no point in repeating the same cost in the costing of each scheme in the comparison. As the main concern is with differences between options, any variables that are common to proposals may be ignored, because they cancel out. All options may then presented in summary table for a report. A typical layout is illustrated in Table 12.5.

From Table 12.5 it can be seen that the option with the lowest construction cost, Option 1, does not provide the best value for money. Although Option 2 has the highest annual equivalent value, when all costs are taken into account, the third option turns out to be the most cost effective building.

Gross development value

The gross development value of a completed building is equivalent to its expected selling price. This selling price is, in turn, equivalent to the discounted value of all future rents less the landlord's costs, discounted at a rate acceptable to a purchaser taking into account interest rates, inflation and the selling price of the property at some point in the future. One method of estimating the gross development value is to find the present value of all future revenues less costs. However, this is seen by many to be too simplified, since it does not take sufficiently into account property market conditions, accounting practices and the objectives of developers. Members of the surveying profession calculate the gross development value using a number of techniques. These will be discussed in turn and each followed by an example of their application.

After many years, buildings usually become obsolete, for many different reasons. Towards the end of their economic life buildings may lie empty or their tenants might be unable or unwilling to pay a high enough rent to maintain the fabric of the structure. Obsolete buildings often need to be demolished and replaced. A separate fund, known as a *sinking fund*, is needed to pay for replacement. Annual payments into a sinking fund ensure that replacements can be financed. This fund is an additional annual cost deducted from annual revenues. It is equivalent to depreciation and compensates for capital consumption. Otherwise, the capital invested in the original building is used up by the time the structure is demolished.

The principle of a sinking fund is that the annual amount set aside for replacement earns interest until it is needed. The sinking fund grows partly because additional money is invested annually and partly because money already in the fund is accumulating interest. For this reason the annual amount needed to provide a given sum in the future is less than the future sum divided by the number of annual contributions. The fund is said to reach maturity once the capital sum has been accumulated.

The sinking fund factor is given in Equation (12.15):

$$SF = \frac{r}{(1+r)^n - 1} \tag{12.15}$$

where

SF = sinking fund factor
r = expected rate of return of money in sinking fund
n = number of years to maturity

The annual sinking fund is the annual sum needed to replace a capital investment.

$$ASF = \frac{Kr}{(1+r)^n - 1} \tag{12.16}$$

where ASF = annual sinking fund
and K = future capital sum required

Assume: $K = \text{£}1000$
$n = 3 \text{ years}$
$r = 5\%$

Substituting in Equation (12.16),

$$\text{ASF} = \frac{\text{£}1000 \times 0.05}{(1 + 0.05)^3 - 1} \qquad (12.17)$$

$$= \text{£}317.21$$

In Table 12.6 the annual sinking fund at the beginning of Year 1 is £317.21. This earns interest in the first year of £9.52, bringing the total sinking fund to £326.73. At the beginning of Year 2, a second payment of £317.21 is made to the sinking fund, making a accumulating total of £643.93. The fund earns £19.32 in interest during the second year. By the end of the third year, the fund reaches maturity.

Combining Equations (12.6) and (12.16), the annual equivalence, including the annual sinking fund, is given as:

$$A = Kr + \frac{Kr}{(1 + r)^n - 1} \qquad (12.18)$$

The procedure for calculating the value of property is based on the years' purchase, using Equation 12.15:

$$YP = \frac{1}{r_t + SF} \qquad (12.19)$$

where

YP = years' purchase factor
r_t = target rate of return of the project

Table 12.6   Sinking fund

	Year 1			Year 2			Year 3	
	ASF (£)	Interest (£)	(£)	Interest (£)	(£)		Interest (£)	Accumulated total (£)
	317.21	9.52	£326.73					
ASF		**317.21**						
			643.93	19.32	663.25			
ASF					**317.21**			
					980.46		29.41	1 009.88

The amount needed to replace £1 at the end of n years at r rate of interest is given by the formula for the SF factor. The result is then fed into the YP equation to produce a factor. Expected annual revenues less expenses is multiplied by this factor to find the gross development value, which is the value conventionally attributed to the property.

A simple example of the application of years' purchase and sinking funds will clarify the method. Assume a building project costs £50 000 per annum and generates an annual income of £80 000, as shown in Table 12.7. A fund is required to carry out repairs at the end of the fourth year.

Let $r_t = 10\%$ (the target rate of return on the investment)

$r = 3\%$ (the safe rate of return in the sinking fund)

and $n = 4$ years.

Substituting in Equation (12.15):

$$SF = \frac{0.03}{(1.03)^4 - 1} \tag{12.20}$$

$$= \frac{0.03}{1.1255 - 1} \tag{12.21}$$

$$= 0.239027 \tag{12.22}$$

and substituting in Equation (12.19),

$$YP = \frac{1}{0.1 + 0.239027} \tag{12.23}$$

$$= \frac{1}{0.339027} \tag{12.24}$$

$$= 2.9496$$

Table 12.7   Costs, benefits and annual yields

Year	Costs (£)	Revenues (£)	Annual yield (£)
1	50 000	80 000	30 000
2	50 000	80 000	30 000
3	50 000	80 000	30 000
4	50 000	80 000	30 000

The gross development value (GDV) of a property is based on expected annual revenues less costs multiplied by the *YP*, and if the expected annual yield is £30 000, according to Table 12.6, then,

$$\text{if} \quad GDV = A \times YP \tag{12.25}$$

$$A = £30\,000$$

$$\text{and} \quad YP = 2.9496$$

Substituting in Equation (12.26),

$$GDV = £30\,000 \times 2.9496 \tag{12.26}$$

$$= £88\,488 \tag{12.27}$$

Tax

A further adjustment is required to take tax into account. Corporation tax is a tax on company profits. If corporation tax is 25 per cent of profits, then a profit before tax of £1000 is only worth £750 after tax. If the amount of tax is known, Equation (12.26) can be used to find profit after tax. Of course, corporation tax takes the overall profit of a company into account rather than individual projects, some of which may generate losses to be offset against the profits made on other deals. The following example is given purely to illustrate adjustments for tax:

$$\pi - t = p \tag{12.28}$$

where

$$\pi = \text{profit before tax}$$
$$t = \text{tax}$$
$$p = \text{profit after tax}$$

Alternatively, the adjustment for tax can be calculated using the tax rate expressed as decimal fraction:

$$p = \pi(1 - t) \tag{12.29}$$

If $\pi = £1000$
and $t = 25\%$ or 0.25

Substituting in Equation (12.29),

$$p = £1000 \times (1 - 0.25) \tag{12.30}$$
$$= £1000 \times 0.75 \tag{12.31}$$
$$= £750 \tag{12.32}$$

Firms set target figures for the amount that they hope to receive after tax. They are therefore interested in the profit figure net of taxes and must calculate the gross figure needed to allow for tax. Equation (12.33) is derived from Equation (12.29). Given a tax rate and the amount required by the developer after tax, the amount needed before tax can be found using the new formula in Equation (12.33):

$$\pi = \frac{p}{(1 - t)} \tag{12.33}$$

This tax adjustment is also inserted in the formulae for the years' purchase and the sinking fund.

The formula for the years' purchase becomes:

$$YP = \frac{1}{\dfrac{r_t + SF}{(1 - t)}} \tag{12.34}$$

At a tax rate of 25 per cent, the value of the above property becomes:

$$YP = \frac{1}{\dfrac{0.1 + 0.239027}{-0.25}} \tag{12.35}$$

$$= \frac{1}{\dfrac{0.339027}{0.75}} \tag{12.36}$$

$$= \frac{2.296}{0.75} \tag{12.37}$$

$$= 3.9328 \tag{12.38}$$

The gross development value of the property illustrated in Table 12.6 can now be calculated, starting from the point of view of the developer who needs to know the price at which the property must be sold allowing for profit after tax. Substituting in Equation (12.25), the annual amount of £30 000 and the tax-adjusted *YP* of 2.21221,

$$GDV = £30\,000 \times 3.9328 \tag{12.39}$$
$$= £117\,984 \tag{12.40}$$

The tax system tends to favour projects with higher running costs and lower initial capital outlays. Apart from private housing landlords, running costs may be set against profits while capital costs may only partly be offset.

Regional variations

The valuation of a site may be based on a comparison with a similar building in a different location built a few years earlier. To make the comparison more realistic, adjustments must be made for inflation and location. As prices vary from region to region, it is necessary to weight prices by using location factors, which can be found in price books, such as Spons and Laxtons, journals, and from the RICS Building Cost Information Service (BCIS). Table 12.8 for example is taken from the *BCIS Quarterly Review of Building Prices*.

To adjust the figures from one region to another, the following ratio is used:

$$\frac{\text{Building price in Region A}}{\text{Region A Location Factor}}$$
$$= \frac{\text{Building price in Region B}}{\text{Region B Location Factor}} \tag{12.41}$$

$$\text{Building price in Region A}$$
$$= \frac{\text{Building price in Reg. B} \times \text{Reg. A Location Factor}}{\text{Region B Location Factor}} \tag{12.42}$$

Inflation

Because of the difficulty of predicting the rate of inflation, it is often assumed to be zero in feasibility studies dealing with long-term cash flows. This may be justified on the simplifying assumption that inflation affects both costs and revenues equally, and the higher prices will cancel each other out. However, if inflation is to be taken into account it is necessary to adjust figures to allow for the effect of inflation on prices. The use of index numbers, such as the Retail Price Index, the Tender Price Index and the Building Cost Index, applies mainly to comparing current prices to past events or prices. However, indices of past performance are not necessarily good indicators of future trends, when

Table 12.8   Regional variation factors based on national average = 1.00

Region	Variation factor
Greater London	1.17
East Anglia	0.99
East Midlands	0.92
Northern	0.96
North West	0.99
Northern Ireland	0.76
Scotland	1.01
South East (excl. Greater London)	1.05
South West	0.96
Wales	0.93
West Midlands	0.96
Yorkshire and Humberside	0.93

Source: BCIS Quarterly Review of Building Prices, March 1996, Issue No. 61, BCIS, Kingston.

changes in government policies or international financial market conditions can be expected to influence the future rate of inflation.

Nevertheless, if the future rate of inflation is assumed to be 10 per cent per annum, then £100 today will purchase the same as £110 in one year's time. The purchasing power of £100 in one year's time is only equivalent to £90.90 today, ignoring the time value of money. The inflation factor is the same as the compound interest rate factor and can be used to calculate expected prices at some point in the future. The inflator is given in Equation (12.43):

$$\text{Inflation factor} = (1 + R)^I \tag{12.43}$$

where R = rate of inflation per interval, (usually a year), and I = number of intervals.

The inverse of the inflator is the deflator, which gives the current equivalent of a future expected price:

$$\text{the deflator} = \frac{1}{\text{Inflation factor}} \tag{12.44}$$

Assuming a rate of inflation of 10 per cent over one year, and substituting in Equation (12.43),

$$\text{the inflation factor} = (1 + 0.1)^1 = 1.1 \tag{12.45}$$

Substituting the inflation factor of 1.1 in Equation (12.44),

$$\text{Deflator} = \frac{1}{1.1} = 0.909 \tag{12.46}$$

Multiplying £100 by the inflation factor, 1.1, gives in the first case the future expected price of £110, while multiplying a future expected price of £100 by the deflator, 0.909, gives the constant price equivalent of £90.90.

These adjustments to future prices may be necessary where rising prices have been given because of anticipated higher labour and material costs or anticipated rent increases, especially when they are expected to be different from the rate of increase in other prices. It would be necessary to deflate such figures to counteract the effect of inflation on the figures. Higher figures in future years in cash flows should only reflect higher real costs or higher real benefits. After prices have been adjusted for inflation, discounting techniques may be applied in the usual way. Table 12.9 presents future sums, expected inflation, deflators and deflated figures or current price equivalents, assuming payment at year end.

Strictly speaking, the real rate of return is:

$$r = \frac{1 + \text{rate of return}}{1 + \text{rate of inflation}} - 1 \tag{12.47}$$

where $r = $ the real rate of return.

Thus, if the rate of return = 14%
and the rate of inflation   = 8%,
then:

$$r = \frac{1 + 0.14}{1 + 0.08} - 1 \tag{12.48}$$

$$r = \frac{1.14}{1.08} - 1 \tag{12.49}$$

$$= 1.056 - 1 \tag{12.50}$$

$$= 0.056 \tag{12.51}$$

$$= 5.6\% \tag{12.52}$$

Residual valuation

In the following example, Parfitt (1987) illustrates a practical application of some of the techniques described above. The object of the exercise is to

Table 12.9   *The deflator factor: anticipated rate of inflation, 10% Year 1 construction cost, £1 000 000 Year 2 construction cost, £1 500 000*

Years hence	Cost (£)	Inflation factor	Deflator	Deflated cost (£)
1	1 000 000	1.10	0.9090	909 000
2	1 500 000	1.10^2	0.8264	1 239 600

ascertain the maximum bid for building land at a site in London for a proposed sports centre, by using residual valuation to establish the value of the site to the client.

The building costs are based on a cost analysis for a light industrial warehouse building, with similar characteristics to the proposed sports centre building. Cost analyses of completed buildings are collected by the Building Cost Information Service (BCIS), which was set up by the RICS. In practice, some adjustments need to be made to the published cost analysis for variations in the buildings.

Development: Light industrial/warehouse at a plot ratio of 0.5.
Proposed development: Sports centre.
Site area: 9500 sq.m.
Gross floor area of proposed development: 4750 sq.m.
Rentals in area for similar accommodation: £59.18 per sq.m
    (full repairing and insurance lease).

Projected building cost based upon an adjusted cost analysis:

Cost analysis:
Industrial/warehouse development in Hampshire, Oct. 1985.
Tender Price Index (TPI), Oct. 1985: 252.
Location factor for Hampshire: 0.99.
Price per sq.m: £221.

Adjustments:
TPI for 1st quarter 1988: 284.
Location factor for London: 1.18.

Location adjustment using Equation (12.42):

$$\frac{1.18}{0.99} \times £221 = £263 \qquad (12.53)$$

Adjustment for inflation:

$$£263 \times \frac{284}{252} = £296 \tag{12.54}$$

£296 is the adjusted price per sq.m. of the site in Hampshire made comparable to the current proposal, by allowing for both the different location and higher current prices caused by inflation.

Assume, for the sake of this example, 5.5 per cent investment yield and 20 per cent developer's profit. An investment yield of 5.5 per cent implies a years' purchase factor of 18.18, since:

$$\text{Years' purchase} = 1/\text{Yield} \tag{12.55}$$

Since 5.5 per cent is equivalent to 11/200, substituting 11/200 in Equation (12.55),

$$\text{Years' purchase} = \frac{1}{(11/200)} \tag{12.56}$$

$$= \frac{1}{11} \times 200 \tag{12.57}$$

$$= 18.18$$

Table 12.10   Site valuation

	£	£
Gross development value		
Anticipated net rental income:		
4 750 sq.m. × £59.18 (p.a.)	281 105	
Years purchase in perpetuity: 18.2		
Gross development value: £281 105 × 18.2	**5 116 111**	
Development Cost:		
Building costs: 4 750 sq.m. × £296	1 406 000	
Architect's and quantity surveys fees:		
10% of building cost	140 600	
		1 546 600
Finance costs: 12% p.a. of costs for 6 months of		
building period £1 546 600 × 12%/2		92 796
Developer's profit: 20% of gross development value		1 023 222
Letting fees (estate agents and solicitors):		
10% of initial rent		28 110
Total cost (excluding cost of site)		**2 690 728**
Site Value = gross development value − total costs		**2 425 383**

From these assumptions and adjustments, it is now possible to derive a site value, which the developer would be prepared to offer for the selected site. This is shown in Table 12.10. The value of the site to the client is therefore £2 425 383, accepting the given set of assumptions. In fact the TPI for 1st quarter 1988 was not 284 as predicted. The figure of 284 was only a best estimate or assumption.

Conclusion

The techniques shown in this chapter are widely used and extremely valuable rules of thumb. Nevertheless, the calculations are based on assumptions about the future and are only as good as the eventual accuracy of those assumptions. The broad assumptions about discount rates, about the future or about using approximate average figures are often matters of convenience and from an economic point of view are somewhat arbitrary. For example, one cannot assume realistically that interest rates will remain constant in the future, though, of course, this assumption is often used in the absence of firmer information. Economists attempt to reduce arbitrary assumptions to a minimum and always state explicitly what the assumptions are.

13 Cost–Benefit Analysis

Introduction

Financial and non-financial aspects of feasibility studies were discussed in the previous two chapters. This chapter is concerned with some further issues raised by and dealt with in cost-benefit analyses, including risk and uncertainty, the social rate of return and very long run implications. Finally, the chapter considers the presentation of cost-benefit analysis reports.

Life-cycle of projects

Life-cycle costing considers the expected useful life of a project as this will affect its viability. The cost of maintenance and refurbishment in future years must be taken into account, as well as the cost of construction. Anticipating the life of components helps to predict likely future expenses and when these will be incurred. This enables building owners to budget for future repairs and maintenance.

Costs-in-use or *annual equivalence* examines the expected annual running costs of buildings, taking into account the cost of finance and the expected life of the building and its components. Indeed, the annual equivalent cost is best suited for comparisons of projects with different expected lives, or when the functions of the building have been predetermined or assumed.

The annual expected running cost figures anticipated at the design stage can be used to give the future building or facilities managers target figures consistent with the design and the decision to build. Building owners can use costs in use to calculate a minimum profitable rental for a property.

There may be a trade-off between higher construction costs and lower annual running costs. The Value Added Tax (VAT) rate on new buildings is zero, while VAT on repairs and maintenance is charged currently at 17.5 per cent. This may have design implications, since more spent in the initial stages in terms of improved quality often leads to lower maintenance costs in future years. This can be particularly important to employers, especially if they plan to occupy the completed buildings themselves.

However, the advantage of constructing buildings with low running costs may be diminished, at least in the UK, by a further effect of the British tax system. Since capital expenditure may be used to offset capital gains, it is treated differently from maintenance and running costs, which may be deducted from income or profits in future years. It is often in the interests of the client, for reasons of tax, to reduce capital expenditure in favour of higher running costs.

Although the durability of components, ease of maintenance and energy consumption may make a property easier to let or sell, these factors may be low priorities for a client if a building is planned for use by tenants or if it is to be sold in the near future. In that case, the burden of future costs would be borne by others and the developer may not be able to recoup the extra expense of high-quality finishes or materials, especially when they are hidden, such as insulation or services.

There are a number of problems associated with costs-in-use. First, the method assumes that a building will, in fact, be maintained according to the pattern dictated by the architect at the design stage. However, many building users see building maintenance as one area that can readily be delayed or even cut. Also, the concept of costs-in-use assumes that money will be set aside every year for repair and maintenance requirements in the future. However, either the setting aside of funds or the actual maintenance is often not carried out. Funding for future repairs is usually set against other current demands for finance when a building is due for major overhaul. The needs of the building will have to compete with other calls on a company's finance. There is a further problem in predicting the useful life of components. Obsolete finishes, especially durable ones, may be demolished before their anticipated useful life has expired. This may be the result of a change of use or a change of taste or fashion. Early demolition, of course, invalidates the cash flow predictions.

Balanced economic arguments

Cost–benefit analysis is a systematic approach concerned with raising arguments for and against proposals, quantifying these arguments, where possible, and structuring the case for or against a proposal in such a way that conclusions may be reached. At the same time, a cost–benefit study should help to anticipate problems of construction and maintenance as well as problems which may be experienced by those adversely affected by a scheme. The aim of the analysis is to present a full picture of the pros and cons of a project in order to help a client come to a decision.

In public-sector projects it is seen as equitable to spread costs over the period of the benefits of a scheme. For example, in the past, a public-sector project such as a reservoir was seen as generating benefits for future generations and new people moving into a particular area. Long term finance was used as a method of spreading the cost of construction over the period of the flow of benefits, otherwise the population at the time of construction would be paying in advance for the benefit of future generations, who would not themselves contribute anything to the capital costs. Costs and benefits should therefore be considered over a period of time. In contrast, in the private sector, funding is usually arranged on a much shorter-term basis. Commercial loans may be spread over three to ten years, but rarely longer. Private household mortgages may be extended for twenty-five or even up to thirty years.

By carrying out cost–benefit studies, solutions may be found by pointing out the relative advantages of particular designs or proposals, even after additional expenses have been taken into account. Proposals may range in size from adding a bar or restaurant to a cinema complex through to designing and planning a city centre site. In fact there is no limit on the size of a project that may be considered using cost–benefit analysis.

Environmental issues

Environmental issues are dealt with in two types of cost–benefit study, Environmental impact analysis (EIA) and urban impact analysis (UIA). The first deals with ecological issues such as the impact a proposal is likely to have on flora and fauna. UIA considers similar environmental issues in an urban context and examines the impact of changes to the built environment on the pattern of life in local communities as well as their health and safety. For example, UIA may be used to consider the impact of information technology on travel and the built environment. Similarly, a study of the impact of a new development in a town centre may be viewed in terms of its effect on the local economy, traffic flows, pollution and the quality of life of the local population.

In EIA the problems of the use of scarce resources and the destruction of land caused by the extraction of building materials are brought into the assessment of a project's overall impact on an area. This can include the safety and environmental impacts of materials at every stage of their life-cycle, through processing, construction, in service, demolition and disposal. The Construction Industry Research and Information Association (CIRIA) have produced a series on the environmental impacts of materials, ranging from mineral products and metals to plastics and paints. The damage to Sites of Special Scientific Interest (SSSIs), inhabited by rare species of plants, insects or animals is also a key issue in many environmental impact studies, as well as damage to beauty spots, rural attractions and social amenities, such as country parks. EIA is also concerned with pollution generated during the life of a project, and in some cases, even after the end of a project. For example,

EIAs would take into account the condition of contaminated land following the demolition of industrial buildings. Energy use is also often a major consideration in environmental impact studies.

The issues raised in EIA and UIA depend on the nature of the project considered and the particular questions raised. Just as each project is unique, there is no single approach to investigating the problems that may arise.

Social rate of return

It is important to bear in mind that even in projects that are run by local authorities and charities reasonable rates of return are necessary to show economic, if not financial, viability.

Benefits are units of gain both to the developer and to third parties whose properties or lives are improved by the construction of a new facility. Costs represent the value of resources used as well as the losses incurred by those adversely affected by projects. A net benefit is the value of a gain after all relevant costs have been deducted. The benefits of a theatre restaurant include receipts from the sale of food, convenience for patrons, general atmosphere in the theatre, and the value of the extra theatre seats sold resulting from the added appeal of the in-house facilities. The net benefit of the restaurant is the contribution it makes after the costs of food, staff and other running costs have been taken into account. Moreover, if, as a result of having a restaurant in the theatre, a local restaurant suffers a loss of custom, then the lost business must be deducted to find the true gain, if any, for society, or at least for the local economy. It is possible that one restaurant might simply replace another, with no overall improvement for the community at large. Indeed, there could even be a net social cost.

The actual rate of return used for non-profit-making ventures may be based on a social rate of return (SRR). The SRR takes into account social costs and benefits which are not always based on market prices but are imputed. The SRR is implied by the average rate of return on decisions taken in the long run. It shows the rate of time preference of society in general. In other words, it is the discount rate implied by the investment decisions one generation takes to have benefits available for future generations.

The derivation of the values of costs and benefits throughout a project's life should be given. For example, staff costs should be calculated by multiplying the wages of staff by the number of staff. The benefits of a museum, even when entry is free, should include an estimate of the number of visitors per day multiplied by the number of days and the imputed value of each visit. All significant variables, especially any which may be disputed, should be presented in the report to clients.

Terminal values

An alternative to finding the present value of projects is to calculate the terminal value or the value of projects at some point in the future. Instead of discounting back to the present, compound interest is calculated on all net benefits from the year they arise. The option with the highest terminal value is then chosen.

This method allows the possibility that profits from a particular project may be withdrawn and used to finance other schemes, perhaps with an expected return greater than that offered by banks or other low-risk investments. Such use of funds cannot be taken into account using discounting techniques, which assume that money is discounted at a chosen rate of interest.

Another advantage of the terminal value method is that the terminal value can be discounted back to the present in the usual way to provide a net present value. Doing this will not alter the outcome of decisions, although the net present value would then be based on a more realistic set of assumptions. By using the terminal value method of evaluating projects it is possible to monitor performance against predictions in coming years. On every count, the terminal value method is superior to the conventional system and will give more reliable and useful results.

The continuation principle

According to the continuation principle, a building should survive only as long as, in any given year in the future, the remaining discounted future benefits are greater than the discounted future costs. Rather than discount figures back to the present, costs and benefits are discounted back to a given year in the future. With conventional discounting techniques using present values, present costs are great relative to costs in the future, even when costs in the distant future are large and significant. These future costs should be compared to future benefits. This is particularly relevant for projects like nuclear power stations, mentioned earlier, in which the costs of maintaining the plant securely after its useful life are likely to be great in future years and to continue costing society for many generations.

Risk and uncertainty

It is important to assess the risk associated with different schemes. This may be achieved by examining different scenarios. A scenario is a set of conditions.

Each time a variable is altered, a new scenario is created. This is useful for examining various possibilities, known as *what if* scenarios, such as: what if sales were 10 per cent higher (or lower) than assumed? The new values create a new scenario with a new discounted present value. This exercise may be carried out any number of times. Computers can now be used to generate a large number of scenarios or random simulations. Because of the similarity with the randomness of a roulette wheel, this method is known as the 'Monte Carlo technique'.

In Figure 13.1, the NDPVs of a number of scenarios for a given option are plotted along the horizontal axis to produce a distribution of present values. The vertical axis shows the number (*n*) of scenarios with a given NDPV. If this exercise is conducted a sufficient number of times, the results spread or form a distribution around a mode or peak, showing the most likely range of outcomes, according to the model and its assumptions. The mode is equal to the arithmetic mean or average when the distribution is symmetric.

The degree of spread or variation of results above or below the mode can be measured, using the standard deviation. The standard deviation is the average distance from the mode. The mode of the distribution of simulations in Figure 13.1 is £350 000 and the standard deviation is £150 000. One of the most important features of a normal distribution is that one third of outcomes lie between one standard deviation above and below the mode. This represents the range of most likely values. In Figure 13.1, the most likely outcome of the project therefore lies between £200 000 and £500 000. The graph also shows the risk of a negative NDPV, as the left-hand tail of the distribution lies below zero.

Apart from the uncertainty about the life of the building, other variables also add uncertainty to predictions about future costs and benefits. Taxation is

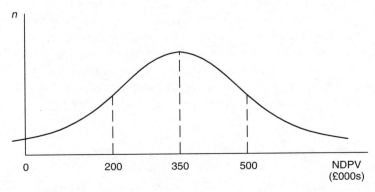

Figure 13.1   Distribution of discounted present values

subject to government policy changes and the financial and economic needs of the country. Inflation has varied dramatically since the 1970s and remains difficult to predict in the long term with any degree of accuracy. Similarly, interest rates have varied greatly over recent years, though it must be added that the variation in interest rates has often reflected changes in the rate of inflation. The variation in real interest rates has not been as great as the variation in current rates. However, a fall in interest rates from 10 per cent to 9 per cent is equivalent to a reduction in the price of money of 10 per cent and can affect cash flow forecasts significantly. Similarly, a rise in interest rates from 4 per cent to 5 per cent is equivalent to a rise in the price of money of 25 per cent.

Establishing true values

The price one pays for a cup of coffee in a restaurant is the price it charges its customers in order to make a profit. However, the price one actually pays is often very different from the value of the actual enjoyment of the drink. This value depends on circumstances, income and tastes. It is quite possible that in some circumstances customers would be willing to pay far in excess of the asking price if they were very thirsty. The value of the drink to the individual would be greater than the price charged. The difference is called *consumer surplus*.

In evaluating a building, the consumer surplus derived from its function and appearance should also be taken into account. This after all measures a value of the building, if not its price. One way of estimating the consumer surplus is to find the compensation value, by asking the public, or those affected, how much they would be willing to accept in compensation if the project did not exist. The equivalent value is found by asking people how much they would be willing to pay in order to have the proposal. It is clear that the answers to the two questions would not produce the same results, but an average of the compensation value and the equivalent value might provide an estimate of consumer surplus. Consumer surplus applies to the amount over and above the price actually paid. It is a measure of benefits not accounted for in the price. Therefore there is no question of double counting a benefit previously measured only on the basis of the price paid.

Incremental analysis

Is extra expenditure justified? Which of two mutually exclusive proposals does one choose? These questions are the kind tackled by looking at the difference to net income a little more or less investment will make. Incremental analysis is used as a financial and economic decision tool.

If an extra £50 000 is spent on a project, it should generate extra benefits. The value of these additional benefits occurring in future years may be discounted back, to calculate the NDPV of the extra £50 000 investment. If the IRR of the proposed addition is above the minimum acceptable rate of return, then the addition of the extra expenditure can be justified, unless there is an alternative proposal. The potential net gains of different options must be compared before making any final decision. These additional gains must outweigh any extra costs incurred over the cost of the next best alternative.

Comparison of two options

Assume that a decision must be taken whether to build new or to refurbish an existing building. A variety of non-financial considerations need to be taken into account. It is usually easier to obtain planning permission to refurbish property than to demolish a building and construct a replacement. Refurbishment is less destructive of the urban fabric, and is often less disruptive than new build, especially to those in the vicinity. However, there can be a greater degree of uncertainty and risk when undertaking refurbishment, though this is by no means always the case. Refurbishment is often used to extend the useful life of a building for a number of years. Consequently, further building work may be required earlier than in new build, though this is not necessarily the case. Some hypothetical discounted costs and benefits associated with two options are shown in Table 13.1.

If refurbishment is carried out, the cost would be £4m, which is the lower cost option. If this option is chosen, net discounted benefits are £2m. If the decision to build new is taken, the cost would be £5m and the net benefit £4m. Thus, in return for the extra expenditure of £1m, the extra benefits are

Table 13.1   Comparison of two options; option 2 preferred

	Refurbishment Option 1 (£m)	New build Option 2 (£m)	Difference Option 2 − Option 1 (£m)
Value of discounted benefits	6	9	3
Discounted total cost	4	5	1
Net benefit	2	4	2

Table 13.2   Comparison of two options, option 1 preferred

	Refurbishment Option 1 (£m)	New Build Option 2 (£m)	Difference Option 2 − Option 1 (£m)
Value of discounted benefits	6	8	2
Discounted total cost	4	7	3
Net benefit	2	1	−1

£3m. By choosing Option 2 in this example, total benefits have been raised by an extra £2m over the next best option.

In Table 13.2 a similar set of figures are presented. However, although Option 2 increases benefits by £2m, the extra cost is £3m, which implies an overall reduction in total benefits of £1m, rather like taking two steps forward and three steps back.

The general layout of cost–benefit studies

A cost–benefit report can be used for a number of different purposes. It can, for example, help architects to promote their designs and defend them against criticism, which ignores consequences implicit in suggestions. It can also enable all types of professional practices to market their services to potential clients. Above all, cost–benefit studies are vital documents for the promotion of any project and an understanding of the implications of any developer's proposals.

Basically, a cost–benefit or feasibility study of a project should present the costs of a set of alternatives, together with recommendations. The recommendations should be based on objective criteria, using discounting techniques. A feasibility study also develops a brief, highlighting possible problem areas, which may require modification to make projects either financially or economically viable. At the early stages of a project the feasibility study is primarily concerned with stating objectives, cost-planning, budgeting and decision-making. Later, the figures may well be revised. This does not reduce the validity and credibility of earlier feasibility studies because, in principle, any change or variation should be adopted only if it improves on an earlier analysis. The modified feasibility study may then form the basis for controlling the project costs and monitoring the operation of the completed building.

Although each study will vary in form depending on the nature, size and complexity of projects, cost–benefit analyses generally consist of several sections. These sections are now dealt with in turn.

Summary and recommendations

A report should begin with a summary and recommendations. The summary should state the recommended proposal, the cost of its construction, the funding required, the period of funding, its IRR and NDPV. There may, of course, be other information deemed necessary for inclusion in the summary. The summary may be the only part of a study actually read by a decision-maker. It is therefore extremely important that it is written with all the information needed for making or justifying a decision, assuming the recommendation is accepted. The length of this part of the report should be kept to a minimum.

Statement of aims

The next section should be a clear statement of the aims of the building, giving the reasons and uses for different spaces within the design. It should also give an indication of the quality the client has requested, an analysis of the number of planned building users, and any other requirements set out in the brief which are relevant for the actual design.

There have been several attempts to help clients to develop their briefs, including Gruneberg and Weight (1990). Latham (1994) notes that many problems in the construction industry arise as a consequence of poor briefing by clients.

Gruneberg and Weight (1990) suggest that each objective should be clearly stated in order of priority. This process may involve resolving disputes and conflicts within the client organisation as different departments compete for resources in the form of space within a new building. Even at this stage, such a discussion helps to focus attention on the aims of a project. From this discussion the design team can develop schemes which take into account the wishes of the client. By delaying a resolution of these issues, the cost of design and construction can increase because designers' time may be wasted on drawings which are never used and, later, partially completed building work may need to be demolished to make way for changes during the construction phase. These are avoidable costs.

Initial and annual costings of preferred option

The report should also contain a section on costings in terms of initial capital and annual running costs. Initial costs may include the cost of site acquisition, construction, and professional fees and finance costs. An allowance may be required for compensation payments to the owners of neighbouring properties, and other contingencies. Annual costs comprise mainly maintenance costs, repairs and services, as well as the staffing required to operate the building, including cleaning costs, which can be extremely high.

Costing of alternatives

A cost–benefit study should also present alternative solutions, giving the client a choice. This allows the firm a second chance if a client rejects the recommended proposal. Alternatives do not need to be presented in as much detail as preferred options but they should be based on a comparable set of figures to show the investment criteria, which a client will require.

Alternative solutions may include refurbishment of an existing building, an alternative shape, size or quality of finish for a building on the same site, another building type on the same site, the same building type on another site, another building type on another site, or a 'do nothing' option. It may be that by reorganising the layout of an existing building or changing working practices no new building is required. Which alternatives are chosen depends on the specific circumstances of the client and the constraints of the brief.

Where costs or benefits are identical for two options, they can be ignored. Thus, if both options use exactly the same site, the cost of the site does not help to establish which is the more viable. Choice is determined by differences, not similarities. Identical costs can be ignored.

Indeed, aspects other than direct costs are often included, such as the quality of construction and finishes, aesthetic qualities, and the comfort and convenience provided by different options. Hedonic pricing refers to the fact that the price of a product is paid in return for a bundle of benefits. Thus a car may be purchased not only because it is a mode of transport from A to B but also because it confers status on the driver, and inclusive features may range from a radio/CD player to air conditioning and a large number of technical features. Similarly, the various features of a building need to be taken into account when considering costs. These factors may be considered in this section.

Financial analysis

A financial analysis should therefore contain at least some of the following information on each option: the capital cost and the capital required, the internal rate of return (IRR), the social rate of return (SRR), the net discounted present value (NDPV), a cost–benefit ratio, annual equivalent costs, and payback period.

The cost of finance (interest payments) may also be included in a financial statement but is not always required in economic analyses. As far as financial arrangements are concerned, the economic analysis considers that the total net benefits from a scheme are divided arbitrarily between the lender and the client, depending on their personal ability to negotiate terms and market conditions. For a commercial project to be economic, it would be expected to generate sufficient benefits, including monetary returns, to meet its financial requirements. This also applies to those projects which require subsidising, since the subsidies would only be forthcoming if the public authority concerned, or the funding charity, considered the economic benefits to be worthwhile.

An elemental cost breakdown

A section showing an elemental cost breakdown of the preferred option would demonstrate to the client the apportionment of costs throughout the design and the derivation of the costs of construction. An element is a functional part of a building, for example, the floors, walls or roof. The total cost of construction in this section should be consistent with the figure supplied in the costing of the cost–benefit analysis.

The categories of elements and sub-elements used are:

1. The substructure, including the sub-elements of site preparation and foundations.
2. The superstructure, broken down into sub-elements, including external walls, windows and doors.
3. The internal finishes.
4. The internal fittings, including furnishing and machinery.
5. Services.
6. External works such as landscaping.

During construction it will be necessary to monitor any changes in costs, to measure deviations from the budget as soon as possible. Variances in the difference between budgeted and actual costs can easily be calculated. See Table 13.3 for an example.

The elemental cost breakdown therefore resembles an approximate bill of quantities, as prepared by quantity surveyors. It may also be useful to show costs as a percentage of total costs for the purpose of comparison with similar projects. Useful reference works for pricing elemental cost breakdowns are the Spon's *Architects' and Builders' Price Book* and Laxton's *National Building Price Book*, as well as past copies of the *Architects' Journal*.

Table 13.3 Cost variance

Element	Initial cost target (£)	Revised cost (£)	Variance (£)	Total variance (£)
External doors	Initial cost target	Revised cost	Variance	Total variance
External doors	3093	2900	+ 193	+ 193
Cladding	50 000	60 000	− 10 000	− 9 807

Conclusion

Carrying out sophisticated economic analyses and financial projections assumes rigorous control over costs during the construction phase and beyond. For this reason, accurate reporting of costs is needed with the minimum of delay, especially during the construction period, when it may become necessary to alter plans. Everybody involved in the design and management of a project needs to understand cost implications, in order to control budgets and spending.

Every project appraisal makes certain assumptions concerning the use of proposed buildings. These assumptions may be used by facilities managers as targets or benchmarks to avoid the failure of the building by achieving the anticipated rates of return on the capital invested. It is *as if* the building were a tool and the feasibility study the set of instructions from the designer to aid the users to get the best out of the building or the project. This applies whether funding is coming from the public sector or the private sector or even coming from the government's strategies involving the Private Finance Initiative (PFI).

Feasibility studies can also be used in post-project appraisal. By comparing the planned scheme with the actual outcome, lessons can be learned and fed into future projects at their planning stages. In this way, mistakes can be avoided and improvements made in project management. According to Gulliver (1987), post-project appraisal highlights the need to determine costs accurately, carry out full market surveys, evaluate key staff, who must remain on the project until completion, and identify and account for problems arising in the course of the design and construction phases.

This book has attempted to give an insight into some of the economic factors which influence relations between firms within construction and between firms and employers. At a time of increasing competition, the survivors in this battle will be those who run their companies and professional practices by demonstrating that management techniques and technology used in other parts of the economy have been applied successfully in their own firms and can be applied equally well by them to the projects they undertake.

Bibliography

Balchin, P. N., G. H. Bull and J. L. Kieve (1995) *Urban Land Economics and Public Policy,* 5th edn (London: Macmillan).

Ball, M. (1988) *Rebuilding Construction: Economic Change in the British Construction Industry* (London: Routledge).

Barras, R. and D. Ferguson (1987) 'Dynamic Modelling of the Building Cycle: 2 Empirical Results', *Environment and Planning A,* vol. 19, p. 518.

Bartlett International Summer School (annual) *The Production of the Built Environment* (London: University College London).

Baum, A. and D. Mackmin (1989) *The Income Approach to Property Valuation,* 3rd edn (London: Routledge).

Bennett, P. H. P. (1981) *Architectural Practice and Procedure* (London: Batsford).

Bowles, S. and R. Edwards (1993) *Understanding Capitalism,* 2nd edn (London: Harper Collins).

Bowley, M. (1966) *The British Building Industry: Four Studies in Response and Resistance to Change* (Cambridge University Press).

Brandon, P. S. and J. A. Powell (1984) *Quality and Profit in Building Design* (London: E. & F. N. Spon).

Bromwich, M. (1976) *The Economics of Capital Budgeting* (Harmondsworth: Penguin).

Browning, P. (1983) *Economic Images* (London: Longman).

Browning, P. (1986) *The Treasury and Economic Policy 1964–1985* (London: Longman).

Butler, J.T. (1988) *Elements of Administration for Building Students,* 4th edn (Cheltenham: Stanley Thornes).

Carpenter, J. B. G. (1982) 'Why Build Faster? – Commercial Pressures', Paper presented to Annual Conference, J. L. O., Oxford, *Building Faster in Britain,* 18–19 September, p. 19.

Cartlidge, D. P. and I. N. Mehrtens (1982) *Practical Cost Planning: A Guide for Surveyors and Architects* (London: Hutchinson).

Chiang, A. C. (1974) *Fundamental Methods of Mathematical Economics,* 2nd edn (New York: McGraw-Hill).

Clarke, L. (1992) *The Building Labour Process,* Occasional Paper No. 50 (Englemere: CIOB).

Coase, R. H. (1960) 'The Problem of Social Cost', *Journal of Law and Economics,* Vol. 3, October, pp. 1–44..

Colvin, H. (1986) 'The Beginnings of the Architectural Profession in Scotland', *Architectural History,* vol. 29.

Computing (1995) 6 April, p. 19.

Cooper, C. (1981) *Economic Evaluation and the Environment* (Sevenoaks: Hodder & Stoughton).

Coxe, W. (1980) *Managing Architectural and Engineering Practice* (Chichester: John Wiley).

Davis, Langdon & Everest, Chartered Quantity Surveyors, UK (1996) *Spon's Architects' and Builders' Price Book*, 121st edn (London: E. and F. N. Spon).

Department of the Environment (Annual) *Housing and Construction Statistics* (London: HMSO).

Derricott, R. and S. S. Chissick (eds) (1982) *Rebuild* (Chichester: John Wiley).

Dornbusch, R. and S. Fischer (1987) *Macro-economics*, 4th edn (New York: McGraw-Hill).

Druker, J. and G. White (1995) 'Misunderstood and Undervalued? Personnel Management in Construction', *Human Resource Management Journal*, vol. 3, no. 3, pp. 77–91..

Farrow, J. J. (1993) *Tendering – An Applied Science* (Englemere: Chartered Institute of Building).

Ferry, D. J. and P. S. Brandon (1991) *Cost Planning of Buildings*, 6th edn (Oxford: BSP Professional).

Gordon, A. (1982) *Economics and Social Policy* (London: Martin Robertson).

Green, R. (1994) *Architect's Guide to Running a Job*, 5th edn (London: Butterworth).

Griffiths, A. and S. Wall (1991) *Applied Economics*, 4th edn (London: Longman).

Gruneberg, S. and D. Weight (1990) *Feasibility Studies in Construction* (London: Mitchells).

Gruneberg, S. L. (ed.) (1996) *Responding to Latham* (Englemere: CIOB).

Gulliver, F. R. (1987) 'Post-Project Appraisal Pays', *Harvard Business Review*, no. 2 April–June, pp. 128–32.

Hallett, G. (1979) *Urban Land Economics* (London: Macmillan).

Hardwick, P., B. Khan and J. Langmead (1994) *An Introduction to Modern Economics*, 4th edn (London: Longman).

Harrison, W. (1986) *Stage One Financial Accounting*, 2nd edn (Worcester: Northwick).

Hillebrandt, P. M. (1984) *Analysis of the British Construction Industry* (London: Macmillan).

Hillebrandt, P. M. (1985) *Economic Theory and the Construction Industry*, 2nd edn (London: Macmillan).

Hillebrandt, P. M. and P. Cannon (1990) *The Modern Construction Firm* (London: Macmillan).

Hillebrandt, P. M. and P. Cannon (eds) (1989) *The Management of Construction Firms* (London: Macmillan).

Inter Company Comparisons (ICC) *Business Ratio Reports* (Hampton: ICC Business Ratios Ltd).

Ive, G. (1983) *Capacity and Response to Demand of the House Building Industry* (London: UCL Press).

James, B. G. (1984) *Business Wargames* (London: Abacus).

Janssen, J. (1991) 'The Development of the Construction Labour Process in Germany' in L. Clarke ed., *The Development of Wage forms in the European Construction Industry* (Dortmund, Fachhochschule Dortmund).

Johnson, V. B. & Partners (1996) *Laxton's Building Price Book 1996: Major and Small Works*, 168th edn (East Grinstead: Reed Information Services).

Keynes, J. M. (1936) *General Theory of Employment, Interest and Money*, 1st edn (London: Macmillan).

Khosrowshahi, F. (1991) 'Simulation of Expenditure Patterns of Construction Projects', *Construction Management and Economics*, vol. 9, pp. 113–32.

Latham, M. (1994) *Constructing the Team* (London: HMSO).

Lawson, M. K. (1987) *Going for Growth, A Guide for Corporate Strategy* (The Latham Reprint, Moores & Rowland).

Layard, R. and S. Glaister (1994) *Cost–Benefit Analysis* (Cambridge University Press).

Legge, R. G. (1935) *Builders' Accounts and Costs* (London: Pitman).

Lewis, J. P. (1965) *Building Cycles and Britain's Growth* (London: Macmillan).

Lindley, D. V. (1985) *Making Decisions*, 2nd edn (Chichester: John Wiley).

Lipsey, R. G. and K. A. Chrystal (1995) *An Introduction to Positive Economics*, 8th edn (Oxford University Press).

McDonald, F. E. (1987) 'Contestable Markets – A New Ideal Model?', *Economics*, vol. XXIII, pt. 1, no. 97, Spring, p. 183.

Mills, E. D. and D. A. E. Lansdell (1985) *House's Guide to the Construction Industry* (London: Van Nostrand Reinhold).

Mishan, E. J. (1988) *Cost–Benefit Analysis*, 4th edn (London: Routledge).

Morishima, M. (1984) *The Economics of Industrial Society* (Cambridge University Press).

Neale A. and C. Haslam (1994) *Economics in a Business Context*, 2nd edn (London: Chapman & Hall).

NEDO (1978) *How Flexible is Construction?* (London: HMSO).

Oxley, R. and J. Poskitt (1996) *Management Techniques Applied to the Construction Industry*, 5th edn (Oxford: Blackwell Science).

Parfitt, T. (1987) *Final Year Architecture Diploma Thesis*, The Polytechnic of North London, (unpublished).

Pearce, D., A. Markandya and E. B. Barbier (1989) *Blueprint for a Green Economy* (London: Earthscan).

Pilcher, R. (1973) *Appraisal and Control of Project Costs* (Maidenhead: McGraw-Hill).

Powell, C. G. (1982) *An Economic History of the British Building Industry 1815–1979* (London: Methuen).

Prest, A. R., D. J. Coppock and M. J. Artis (1994) *The UK Economy*, 13th edn (Oxford University Press).

Rougvie, A. (1987) *Project Evaluation and Development* (London: Mitchell).

Ruddock, L. (1992) *Economics for Construction and Property* (London: Edward Arnold).

Schwartz, M. (1986) 'The Nature and Scope of Contestability Theory', *Oxford Economic Papers*, vol. 38, November.

Seeley, I. H. (1996) *Building Economics*, 4th edn (London: Macmillan).

Squire, L. and H. G. van der Tak (1975) *Economic Analysis of Projects* (Baltimore, Md: Johns Hopkins University Press).

Stone, P. A. (1983) *Building Economy*, 3rd edn (Oxford: Pergamon Press).

Stone, P. A. (1988) *Development and Planning Economy* (London: E. and F. N. Spon).

Treasury (1994) *The Budget in Brief* (London: HMSO).

Williamson, O. (1975) *Markets and Hierarchies* (London: Collier-Macmillan).

Woolf, E., S. Tanna and K. Singh (1986) *Systems Analysis and Design* (London: Pitman).

Index